Implementing Flexible Manufacturing Systems

Implementing Flexible Manufacturing Systems

Nigel R. Greenwood

A HALSTED PRESS BOOK

JOHN WILEY & SONS
New York

First published 1988 by

MACMILLAN EDUCATION LTD
London and Basingstoke

Published in the U.S.A. by
Halsted Press, a division of
John Wiley & Sons, Inc., New York

Printed in Hong Kong

ISBN 0–470–20932–1

Library of Congress Cataloging-in-Publication Data
Greenwood, Nigel R., 1953–
    Implementing flexible manufacturing systems.
    "A Halsted Press book."
    Bibliography: p.
    1. Flexible manufacturing systems. I. Title.
TS155.6.G74   1988   670.42   87–14969
ISBN 0–470–20932–1

*To my mother,
and in fond memory of my father*

Contents

Foreword

All too occasionally an internationally regarded expert finds the time and motivation to fill a glaring gap in the texts written on his subject. The author must be very highly committed to his task, since the effort involved is substantial and brings an additional 'avoidable' workload to an already busy lifestyle.

I believe we are fortunate in that this book is the product of such a rare event.

The subject of Flexible Manufacturing Systems (FMS) could not be more apposite to the future of manufacturing industry. FMS theory and practice is in my view quite simply the single most generic and practicable perspective from which to view the next two decades of manufacturing systems development. Despite its central importance, however, the subject has not seen the adequate publication of much needed quality texts.

Perhaps the particular breadth and depth of understanding required to write convincingly in this area has deterred otherwise eager technical journalists. Whatever the reason for the dearth of quality books on FMS, I am delighted to see this thoughtfully prepared volume arriving to fill the void. The successive chapters deal evenly and objectively with concepts, pros and cons, and go on to provide an excellent narrative of guidance for system designers and implementors, as well as those charged with relevant management decision-making.

Dr Greenwood's text clearly contains much of his accumulated knowledge on the subject of FMS and is immediately recognisable as an authoritative statement. I hope and believe that the distinction will not prevent the book from gaining the very wide audience that it and the subject of FMS truly deserve.

Keith Rathmill

Preface

The purpose of this book is to provide both an introduction to the concept of flexible manufacturing and guidance for those embarking on, or considering embarking on, the implementation of such a project.

It should be appreciated that this book is about manufacturing strategies, rather than automation. It is not an attempt to answer all the questions likely to confront the enthusiastic FMS designer. This would be too bold an objective. Instead it is an attempt to raise the major issues. There is much expertise and literature available to answer application-specific questions, and a comprehensive bibliography is included in References and Recommended Reading at the end of the book.

The firm belief of the author is that flexible manufacturing systems, in one form or another, represent how the majority of manufacturing will be carried out in the future. Soon, a production engineer will not be considered experienced without some exposure to FMS. However, while technology has advanced to the point where the technical risks associated with FMS have been substantially reduced, it is still not easy to implement such systems successfully. This is due to the influence of a variety of factors, many of which are discussed.

It is hoped that through this book the reader will be led to consider issues which might otherwise have been overlooked. The potential benefits which can be reaped as a result of implementing an FMS successfully are immense. However, when embarking on an FMS project, one should not:

(1) Underestimate the long-term impact of installing the system.
(2) Underestimate the difficulty of finding the right people and technology to implement the system successfully.
(3) Underestimate the training and safety implications.
(4) Be constrained by traditional approaches to manufacturing when designing the system.
(5) Underestimate the amount of work.
(6) Be hasty in choosing one's consultants and suppliers.

It has been said that life is not really full of problems, merely challenges. If this is the case, then FMS is one of the most exciting and potentially rewarding challenges facing manufacturing system engineers.

I hope that both managers and students interested in flexible manufacturing will find this book of value, not necessarily because it answers all the questions, but because it is a legitimate attempt to raise what generally appear to be the most important questions.

Nigel R. Greenwood

Acknowledgements

This text would not be complete without an expression of my gratitude to the following people and organisations whose generous help made this book possible:

Gordon Mackie

Bernard Carey

Mel Chudzik

Andrea Hill

Mike Jeffries

Ralph Patsfall

Prakash Rao

Keith Rathmill

Sir Jack Wellings

Deckel Machine Tools GMBH

Fanuc Ltd

Frost and Sullivan

General Electric Company (USA)

Eaton-Kenway Inc.

LTV Aerospace & Defence Company

Scamp Systems Ltd

S.N.E.C.M.A.

Okuma Machinery Ltd

Werner Kolb

Not forgetting Macmillan Education for their seemingly endless patience.

1 An Introduction to Flexible Manufacturing

1.1 What is flexible manufacturing?

Probably the best way to start a book such as this would be with a concise clear-cut definition of flexible manufacturing; unfortunately, this is not going to be possible. There is such a plethora of definitions, not to mention an even greater number of different interpretations, that what might appear to be a simple task is in fact far from straightforward. A situation which is somewhat analogous to that of FMS itself.

Although flexible manufacturing is the manufacturing topic that has been discussed and written about most in the various trade journals and conferences over the past few years, the problem of a suitable definition has yet to be solved. In fact, if anything the issue of definitions is becoming even more complex — with the discussions now concerning the related subject of computer integrated manufacturing (CIM).

Certainly, everybody seems to agree that one of the main purposes of investing in an FMS is to bring the economies of scale of mass production to small batch production. This is a transition which is long overdue, since in the industrialised world small batch production accounts for a substantial proportion of most nations' gross national product. However, despite this, for the past 75 years the majority of manufacturing research and the subsequent advances have been devoted to improving the efficiency of the mass production sector, which is admittedly an influential market sector, but probably not the most important. Before selecting a definition of flexible manufacturing for the purposes of this book, it is worthwhile considering the origins of the technology.

1.2 The origins of flexible manufacturing

The origins of flexible manufacturing technology lie in the desire to automate small batch manufacturing (that is, the production of parts in batches generally smaller than 50). The relative importance of this type of manufacturing is surprisingly high, as is shown in figure 1.1. Within most industrialised nations approximately 30 per cent of Gross National Product is represented

1

by manufacturing. Of this, 40 per cent is batch manufacturing and only 15 per cent mass production. Of the batch manufacturing, 75 per cent of this is in batches of less than 50. Typically, therefore, about 10 per cent of most industrialised nations' GNP is devoted to small batch production.

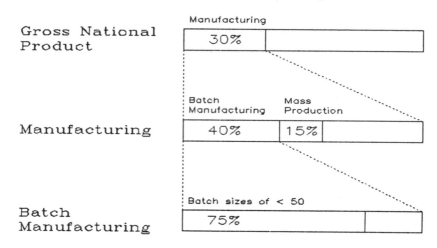

Figure 1.1   The significance of small batch manufacturing

Previously, it was felt that the only way to produce high-quality products quickly and economically was by mass production. During the days of the Industrial Revolution, and for some time after, this was probably true. However, technology has now advanced to the point where there is a viable alternative, namely 'flexible manufacturing'. However, currently this is still a relatively new approach to manufacturing, and hence continues to be subjected to the many difficulties associated with the application of all new technologies.

The predisposition towards mass production is in some respects surprising, since the resultant manufacturing process is plagued with complex-system-oriented inefficiencies, such as complex scheduling requirements, lengthy set-ups, expensive tooling and excessive work in progress.

Within a typical, essentially manual, small batch manufacturing environment, the situation is in some respects worse. Basic production processes, such as machining, are relatively slow because automatic equipment is neither available nor justified, and the skilled labour element cannot be added quickly. Nevertheless, the operating overhead of such organisations remains low when compared with the typical mass production company. This explains why there are often so many jobbing shops in close proximity to high-volume manufacturers; these jobbing shops buffer the mass producers against fluctuations in market demand.

The main reason why small batch manufacturing has, until recently, been left behind in the automation arena, is because the mass producers were, and to some extent are still, so influential, both commercially and technically. They have been able to dictate the type of equipment that was developed, for example, by machine-tool manufacturers. The much smaller, less unified organisations representing the small batch manufacturers have had to be satisfied with the equipment which was built to meet production requirements quite different from their own. However, times are changing.

Recent advances in numerical control (NC) machines have offered batch manufacturers an up-to-date and efficient means of reducing production cost while increasing both productivity and flexibility. However, it is interesting to note that some 95 per cent of machine-tools in most countries are still non-NC. This probably represents some 70–85 per cent of machine-tools by value, well below the generally accepted steady-state value of 50 per cent computer numerical control (CNC), the more advanced form of NC which exists today.

However, when this and other advancements in manufacturing technology are taken together with the consumer market's desire to have more varied products, it is hardly surprising that flexible manufacturing has become extremely important, since it leads to the economic production of a wide variety of parts, with many of the benefits previously associated only with mass production. This, when taken into account with the market demands for increased product flexibility, means that many manufacturers, in order to maintain their market share, must now excel at producing a variety of products rather than just a large quantity of the same product.

Even though this has ensured the continued growth in the popularity of flexible manufacturing, it has nevertheless caused some confusion. This is essentially because, depending on the production environment in which one is operating (for example, mass production, small batch manufacture etc.) the whole concept of 'flexibility' is entirely different. It is now being appreciated that the resulting confusion does not necessarily impact significantly on either the applicability or the philosophical approach that needs to be adopted for the successful application of flexible manufacturing. Indeed, it is not only the disciplined approach to the development of successful FMS which appears to be independent of the manufacturing environment, but much of the technology is also remarkably similar in function, even if it tends to look somewhat different.

It is for these reasons therefore, that for the purposes of this book, the definition which will be used for a flexible manufacturing cell and system (the former simply being a slightly smaller version of the latter) is as follows:

'A *flexible manufacturing system*, through the careful combination of computer control, communications, manufacturing process and related equipment enables a section of the production-oriented aspects of an organisation to respond rapidly and economically, in an integrated manner, to significant changes in its operating environment. Such systems typically comprise: process equipment (for example, machine-tools, assembly stations, robots etc.), material handling equipment (for example, robots, conveyors, automated guided vehicles etc.), a communications system and a sophisticated computer control system.'

Admittedly, this definition is both lengthy and general; however, most definitions of FMS tend to be of limited scope and are compared somewhat subjectively. Usually, they have been evolved with a particular manufacturing requirement in mind, and certainly they have not been able to withstand the test of time. Indeed, there are some other reasons why this definition is perhaps more resilient than many. For example, it is independent of the process technology associated with the system itself. This is a distinct advantage since processes tend to be associated with particular products and production environments, both of which tend to change significantly with time.

Also, some flexible manufacturing systems incorporate a mixture of CNC, NC and automatic cycling machines, not only CNC machines, as many people often assume FMS should comprise. Similarly, in many machining-centre-oriented systems, robots do not appear at all. Most important of all, the whole approach to computer control is changing so rapidly that any attempt to tie in a definition of FMS to a particular computer system architecture would inevitably be doomed to failure.

Furthermore, within this definition the prime motivations for installing FMS are clearly identified; these being the economic manufacture of a large variety of products for which mix and volumes are likely to change significantly, and the reduction of work in progress and throughput times.

Interestingly, this part of the definition actually threatens to exclude some so-called flexible manufacturing systems because, for example, a large product library is not an integral part of their design.

Theoretically, according to a variety of sources, there are some 100–150 flexible manufacturing systems around the world (see figure 1.2). It is doubtful whether all these systems are true FMS as defined above, however they are at worst highly sophisticated automation systems. But the mere fact that virtually all of these have been commissioned during the past few years, gives an indication of the importance that this technology is accorded.

Ultimately, it is important to interpret the definition in terms of the typical operating environment in which the FMS is placed, and perhaps to consider the presence or absence of certain secondary features which tend to distinguish automated systems from flexible manufacturing systems, such as real-time scheduling and recovery capabilities etc.

1.3 The development of FMS

Probably nobody would deny that recently FMS technology has been advancing rapidly, especially now that most of the initial euphoria has died

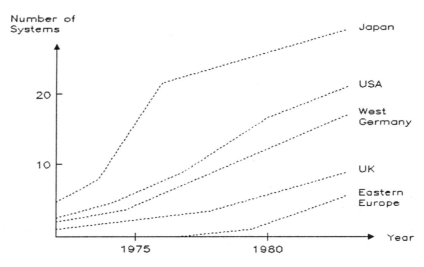

Figure 1.2   The rise in popularity of FMS worldwide

away and been replaced by a combination of commonsense engineering and advanced technology. This has helped to expel the traditional view that FMS was only applicable to a relatively small proportion of the middle of the manufacturing spectrum. In fact, it is now becoming widely accepted that the principles underlying FMS may be applied with equal success within the realms of mass production, and the small jobbing shop (see figure 1.3).

Only recently, probably since about 1980, has the true potential of flexible manufacturing become both viable and appreciated ('viable' in this context meaning technically feasible, reliable and cost-effective). However, there is still considerable confusion about the true meaning of 'cost-effective' when applied to sophisticated automation projects such as FMS. Similarly, few understand the true implications of installing an FMS in an otherwise traditional manufacturing plant. To some extent, this is inevitable; after all, it would be most unusual if such a significant change in methodology was absorbed without causing some conflict.

It is generally understood that FMS is not easy to implement successfully. In fact, it is quite the opposite. This is to be expected, given the extraordinarily complex combination of advanced technologies that have to be both available and correctly applied for an FMS to succeed. So, given the way in which manufacturing technology has been advancing over the past few decades (see figure 1.4), it is easy to see why reliable and economic FMS has been relatively long in arriving.

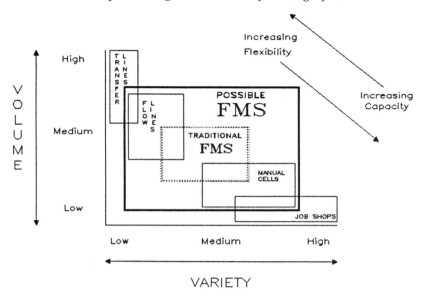

Figure 1.3   The applicability of FMS to manufacturing

Obviously there is a learning curve associated with the development and the application of any new technology. While the technology necessary for FMS certainly exists, there is now the question of developing the expertise necessary to design, implement, operate and maintain these systems success-fully. This is a long-term process which is common to virtually all areas of technology — an initial learning curve, lagged by the application of the development. Admittedly, within different technological environments the timescales will vary substantially, sometimes with the situation being aggra-vated by the fact that the solution to a problem has been developed even though the problem itself has not yet been fully defined. This is generally agreed to have occurred with laser technology, which spent many years waiting for an application, and now has almost too many to be listed.

However, although the technological steps are becoming larger, the subsequent time-lags before application are tending to be reduced in dura-tion. This fact is sometimes obscured by people taking the opportunity to demand increasingly more complex solutions each time. For FMS, it has taken nearly 25 years for the technology to start appearing as a truly viable solution to many companies' manufacturing problems. Indeed, it is interest-ing to consider some of the more important events which have led to the development of FMS.

Perhaps surprisingly, one of the most significant events which led to the birth of FMS was the conception of the Model T Ford. Not that Henry Ford was particularly concerned with manufacturing flexibly (quite the contrary in fact), but because Henry Ford's approach to manufacturing made the

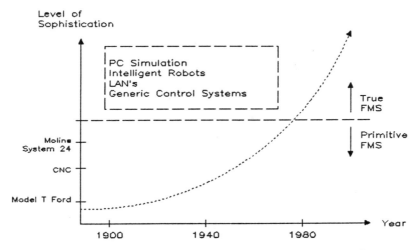

Figure 1.4   The rate of advance of manufacturing technology

development of flexible manufacturing systems almost inevitable, essentially because it represents the opposite of everything that FMS represents, in terms of overall production efficiency and the ability to respond to market pressures.

In 1907 the Model T was born. The launch marked one of the first instances where a large number of accurately machined mechanical components were combined to form a product. The product was sufficiently inexpensive to purchase that it sold in significant volumes (for that time). Indeed, during the sixteen-year life of the Model T, over 19 million vehicles were sold. At the height of production, over a million cars were being manufactured each year. But of course, the Model T, perhaps unjustifiably, gained a certain amount of notoriety, because at one stage Henry Ford was supposed to have said about his brainchild "you can have any colour, so long as it's black. ." (strange to think that nowadays, one usually has to pay more for this option!).

What this meant was, that everyone could obtain the product, relatively cheaply, but there was little or no variety. Your car was precisely the same as that of your neighbours! This is where one of the main advantages of FMS becomes apparent; it accommodates the individual's desire to be truly individual. It makes possible the manufacture of the same basic product at the same price, but with a certain degree of customer-selectable variation. In these days of volatile markets, rapidly changing tastes, fashions, technologies (and hence short product lives), the ability to vary the basic product is likely to prove to be an essential element of industrial survival.

This factor, together with that of the basic importance of small batch manufacturing, provides the fundamental motivation for the development of flexible manufacturing technology. All that was needed for this to become a reality was the development of an appropriate approach and adequate technology.

1.4 Group Technology

In fact, many of the principles on which FMS is based find their origin in research which dates back to the late 1940s and the early 1950s. Of particular importance was the development of the manufacturing philosophy commonly called Group Technology. This effectively uses the common properties of similar parts to solve common design and manufacturing problems. A part coding system is usually used to help classify parts into groups according to, say, their shape, overall dimension, accuracy requirements, surface finish, manufacturing requirements, material composition etc. This analysis is initially applied to all the major components manufactured by a company. The information obtained is then sorted into major categories, such as products or processes, to identify useful production groupings and, for example, groups of machine-tools necessary to manufacture the parts.

These groups, because they are small, distinct manufacturing units, are then usually able to produce the parts more efficiently than if they were being produced as a part of a total manufacturing requirement passing through a large shop. The importance of Group Technology to flexible manufacturing is often overlooked. If maximum benefit is to be derived from an investment in FMS, manufacturing problems must first be sorted into well-defined groups; these would include the processes involved, the equipment required and the parts to be produced. If this analysis is not carried out, it is likely that resources will be directed in a far from optimal manner within the FMS. However, ultimately the production requirements of the eventually selected group of parts should be addressed in their entirety by the equipment within either one FMS or a set of sequential systems. Anything other than this ordered flow leads to a situation not unlike that associated with a functional equipment layout, and its inherent inefficiencies. Indeed, in many plants the application of group technology alone has generated a startling number of benefits.

Within one plant in the USA, components often needed more than a hundred operations, being handled at least at the start and finish of each of these; and during their 5-month average throughput time, they travelled more than 3 miles on the shop-floor. After group technology was applied, the number of operations was reduced from hundreds to tens, with the distance travelled being reduced to 200 feet, and the throughput time to a few days. So successful was this first application that it is now part of a major reorganisation program within the plant. Between June 1986 and December 1986, 125 machine-tools will be moved into groups, affecting some 70 000 square feet of shop-floor.

One of the keys to the success of this program has been the desire of the workforce to become multi-skilled and to play a greater part in the operation of the cells. As such, the number of shop-floor skill levels within the plant has been reduced from 11 to 2.

1.5 Computer numerical control

While group technology was being refined, what is probably the most important technological FMS building block was developed, namely numerical control (NC), which was later to become computer numerical control (CNC).

Numerical control machines were first developed in 1952; in the USA, by the Massachusetts Institute of Technology and, almost at the same time, by Alfred Herbert in the UK. These first continuous-path machines tended to be somewhat unreliable and it was not until 1956 that Ferranti produced a machine of almost acceptable reliability. To a large degree, it was the aerospace industries that instigated these developments, since they had an on-going requirement for complex, frequently 'copied' components, which at that time could only be produced repetitively from experience, using often inaccurate (relatively speaking) hand-made templates. Initially, the computer equipment required to control the machines was large and expensive, but cheaper point-to-point machines were already being developed. As technology progressed so advances were made with the machine-tools and their controllers. Features quickly became more comprehensive, more reliable and less expensive.

As further advancements were made in computer technology and with the advent of complex integrated circuits, true computer numerical control was developed. These controllers, being centred around a small minicomputer, were far more versatile and reliable, and often less expensive than their numerical control predecessors. They also incorporated such features as computer aided programming, sophisticated editing facilities, etc. All of these considerably added to the flexibility of the machine and, for example, the ease with which it could be programmed and set-up to produce a different component. In some respects the culmination of these developments is probably the modern-day machining centre, equipped with an automatic tool changer, tool storage system and automatic work transfer and off-line programming capability.

Although destined to go through a phase of some 15–20 years of developing a reputation for complexity, unreliability and lack of user-friendliness, CNC provided the technological foundations on which FMS could be built. The ability for a computer, instead of an operator, to control a manufacturing process proved to be a fundamental breakthrough. Initially, CNC was only devoted to machining, but now it is applied to a wide variety of manufacturing processes. The continuing evolution of CNC has led to the development of advanced robots. These are now applied to mechanical handling, arc and spot welding, paint spraying and many other applications. This process of broadening the applicability of CNC has been enhanced considerably by the almost unbelievable progress which has been made in computer technology. The combination of the two has also given manufacturing systems designers such tools as programmable logic controllers (PLC), automated guided vehicle systems (AGVS), automated storage and retrieval systems (ASRS) and many other of the standard production engineering building blocks of today.

However, despite all these individual developments, there still remained two significant hurdles to be surmounted with regard to the ability to create sophisticated automated small batch manufacturing systems. First, the integration of all these elements into a reliable working system, and second, the minimisation of the costs and risks associated with once-off control software. Both these issues are now being addressed, and are discussed in some detail in chapter 12.

1.6   The history of FMS

In England during 1968, what is generally regarded as the first flexible manufacturing system was developed by D. T. N. Williamson who was working for the Molins Machine Tool Company. However, it should perhaps be mentioned that some people feel (incorrectly in the author's opinion) that it was Project Tinkertoy, sponsored by the US National Bureau of Standards in 1955 that marked the true birth of FMS.

It is not clear what prompted the development of FMS at this moment in time. It was probably due to the interaction of a number of factors; for example, the progress which had been made with machine-tool technology, the theoretical work on related topics such as job shop scheduling (which had been proving quite successful since its inception during the early 1950s), the success that group technology was having in a variety of different manufacturing environments, and certainly a requirement for being able to produce components economically in small batches was evolving.

At all events, the Molins company which was, and indeed is, a successful manufacturer of high-quality, very high speed machines (notably for the tobacco industry), started to develop what it called 'System 24'. It was designed to produce light flat alloy components, the market for which it was felt was going to develop significantly. The machining system was designed to meet three specific goals:

(1) It had to be capable of manufacturing a large variety of components, virtually at random.
(2) It had to be capable of both loading and unloading tools and workpieces automatically.
(3) It had to be capable of operating, virtually unattended, for long periods.

These goals were indeed very far reaching, and in many ways are still the aims of FMS today. Molins obviously felt this was likely to be the case since some equally far-reaching patents were filed. Within the UK, these have since expired, but within the USA, for various legal reasons, they have only just come into effect. It will be interesting to see whether any of the existing systems and developments contravene these patents.

Unfortunately, for a variety of reasons the Molins System 24 did not succeed. The market for light flat alloy components did not develop as had been predicted, and the machines themselves had included some advanced technology which proved to be quite unreliable (for example, within the control system and the hydrostatic slideways). So eventually the development programme was cancelled, after having cost a substantial amount of money. The overall effect of this was that within the UK, FMS technology virtually ceased to progress for nearly ten years.

But the FMS candle had been lit, particularly in technology-conscious countries such as Japan and the USA. A number of representatives of each country had studied the System 24 development with great interest and, although the UK abandoned the technological lead it had developed, a number of companies, mainly machine-tool manufacturers (as to some extent is still the case) started to build what are now generally regarded as the first, albeit 'primitive', flexible manufacturing systems. These were primitive, not because they were not advanced for their day, but in comparison with the technological developments made since, which have put the capabilities of these earlier systems in perspective. This is apparent if one considers some of the major features that many of these systems had in common:

(1) Process equipment which was not sufficiently flexible to encompass adequately the varied production requirements.
(2) Fundamentally inadequate supporting technologies. For example, many of the sensors used within these early systems were limit switches which, while being notorious for their unreliability, were designed into the FMS in such a way that their failure could stop the whole system. Also, the earlier machine-tool controllers were not designed with communication and integration in mind.
(3) Nearly always the systems were machining-centre-oriented. This was essentially because machining centres had advanced rather more rapidly than most other types of machine-tool, largely as a result of the large work content of the typical cubic component. As such these machines were equipped with reasonably effective control systems, pallet and tool changers etc., all of which accelerated their incorporation into these early flexible manufacturing systems. This in turn biased the direction of FMS.
(4) The systems were often wrongly motivated (for example, purely to take advantage of government grants, or to jump on the technology band-wagon).

Consequently, many people, quite incorrectly came to the conclusion that FMS was only applicable to the manufacture of cubic components: a belief which to some degree persists even today. But, perhaps most significantly, what was missing from many of the early systems, was really what they should have been designed for in the first place, namely, flexibility. Usually, the part libraries of these systems were quite small, and the work necessary either to introduce a new part, or merely to change from manufacturing one set of parts to another, was quite substantial.

Although this phase of the development of FMS technology was a very necessary one, and indeed typical of the evolution of virtually all advanced technologies, it is perhaps unfortunate that many people established their opinions about FMS so soon after its conception. Indeed, this is a story not unlike that which occurred following the development of NC, where people were only too pleased to sit in judgement before allowing sufficient time for the application of the technology to stabilise. It is probably true to say that only recently has production engineering come to grips with the concepts and the technological components of FMS, and hence progressed beyond the formative stage.

1.7   FMS today

It is now becoming relatively straightforward to find examples of reliable 'true' FMS. Several such systems are described in the following chapter. The real legacy of this earlier phase of the development of this technology, and that of Henry Ford, is the general acceptance of the fundamental truth that flexible manufacturing is not only a means of production, it is also, if not more so, a philosophical approach which results in carrying out a set of tasks efficiently. It is an approach that in fact may quite easily be extended to cover far more than just manufacturing. To emphasise this point, the 'traditional approach to batch manufacturing' as in many respects pioneered by Henry Ford and others, may be compared directly to the 'flexible approach to manufacturing' as practised within the more successful and hence more convincing examples of FMS. A summary of such a comparison is shown below:

The traditional approach to batch manufacturing, as found even today within many functionally laid-out factories is

(1) Subdivide jobs into many simple operations.
(2) Complete operations on batches sequentially.
(3) Complete operations very quickly.
(4) Automate individual operations.

The flexible approach to manufacturing is

(1) Subdivide jobs into few operations.

(2) Overlap operations on batches wherever possible.
(3) Complete operations consistently quickly.
(4) Automate entire operation sequences.

In many factories throughout the world even today, production engineers tend to split a particular component's manufacturing process into as many separate operations as seems practicable, ostensibily to facilitate the balancing of equipment utilisation and to simplify the production task. These operations are all completed on an entire batch of parts, completely isolated from the other operations. These individual operations are then completed as quickly as possible, given the capabilities of the production equipment. Any automation is applied to an individual operation, rather than the complete manufacturing process.

This has helped maintain the situation where some 95 per cent of the throughput time for the average workpiece is spent either in storage, transit or simply waiting. Even when the part is on a machine-tool, it is only being 'processed' for about 30 per cent of that time. The result is that, for example, metal removal often only accounts for about 1.5–2.0 per cent of the total production process time for a machined part. Other consequences are, for example, lengthy throughput times, poor machine utilisation, excess work in progress inventory etc. This is obviously extremely inefficient from both the production engineering and financial standpoints. FMS offers the opportunity to improve this situation by reducing many of the inefficiencies associated with these factors.

A somewhat more clandestine result of strong traditions such as these, is the lasting impression that the traditional approach is the only approach worthy of merit. As far as manufacturing was concerned, this was, in some respects, reflected in the process equipment which was developed — for example, machine-tools such as bar automatics. These could produce relatively complex parts extremely quickly, once they had been set-up. Unfortunately the setting-up process took so long that massive batch sizes had to be run ostensibly to enable the facility to be operated economically. Automation advances were almost entirely devoted to the development of this type of machine-tool. Little regard was paid to the fact that their continued use was in fact resulting in components being produced relatively inefficiently. Such a situation rapidly becomes serious when the requirements for parts reduces below that of the economic batch size, since then one has the choice of either producing fewer components than are needed to run the machine economically, or having to pay for the storage of the excess components that are manufactured.

It is generally accepted that within a relatively efficient factory an operation on a batch of components may be processed on one machine in a week. This 'rule of thumb' remains valid for a surprisingly large number of factories in production today. It is indeed a great shame that nobody told Henry Ford that it was likely that the majority of his components were only being machined for a small proportion of the time they were on the shop-floor. He might well have tried to develop FMS earlier.

Compared with the 'traditional approach to batch manufacturing', the 'flexible approach to manufacturing' is appealingly simple. It essentially requires that manufacturing processes should be split into as few operations as possible. This immediately reduces the number of set-ups and inter-operational transfers, thus improving the overall efficiency of the manufacturing process (see figure 1.5). Today, this frequently means that sophisticated production equipment, such as machine-tools with tool and pallet changers, is required to carry out these more complex operations, since once a part is on a machine, as much work as possible must be carried out before the part is taken off. But this type of machine-tool, which is becoming almost normal rather than the exception, is usually particularly well suited to this type of manufacturing environment.

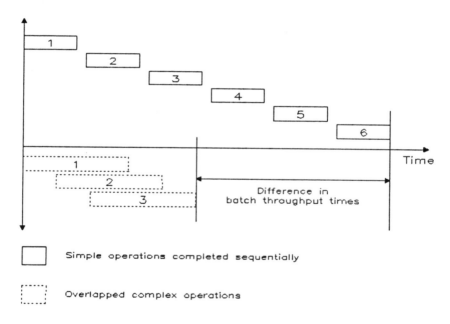

Figure 1.5   The effect that overlapping operations has on throughput time

Also operations should be 'overlapped' wherever possible. Ideally, as soon as one component from a batch has been completed on one operation, it should be passed to the next operation. The machine from which it came is then free to proceed to the next part (which may or may not be from the same batch, as would be necessitated by a traditional mass production system). This results in a dramatic reduction in the throughput time for the whole batch.

Similarly, the individual operations should be completed as quickly as consistency, accuracy and/or quality, and reliability in both the equipment and the process will allow. Since the FMS is likely to be operating within an environment containing few people, it is essential that the system is reliable, and hence not poorly utilising the services of probably the scarcest resource in the system, namely the operators. If, for example, machine-tools keep breaking down because they are being run at either their, or their tooling's maximum capability, it is unlikely that the system as a whole will be particularly efficient. It is far better to settle for a consistent, medium rate of operation. It should be appreciated that the goal of flexible automation is not to complete an equivalent task faster than possible by a human counterpart (for example, the loading of a relatively light component into perhaps a CNC lathe). That is perhaps more the task of what is generally called 'hard automation'. The goal of flexible automation should be to meet a set of components' entire manufacturing requirements (see figure 1.6). This should result in a system which will be capable of reliably producing consistently good parts seven days a week, twenty-four hours a day, if needed.

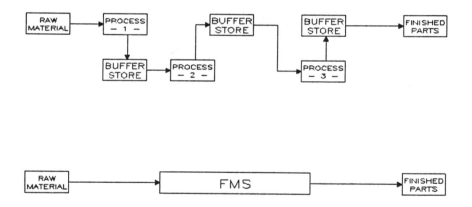

Figure 1.6   Throughput times benefit from automation of an entire process

1.8 The constituents of FMS

Irrespective of the actual environment in which the flexible manufacturing facility is located it is likely to comprise similar functional elements, that are essentially of two basic types:

(1) Workstations — at which workpieces are processed or altered in some way.
(2) Storage units — where work pieces are purely stored and hence do not change in form, such as buffer stores, ASRS etc.

Without doubt there is considerable confusion within production engineering concerning the definition of the elements of an FMS. For example, sometimes a workstation is used to describe an individual machine-tool, perhaps with a tool changer etc. On other occasions a workstation might represent a number of separate, albeit closely related, machines (sequential presses etc.). Such a group of equipment might, by some people, be called a module, or a subsystem or even a cell, which for others (indeed the majority) represents a small manufacturing system.

For the purposes of this text a workstation will be regarded as an individual process device, perhaps manual, or perhaps full CNC with various ancillary equipment, such as toll changers etc. This seems sensible since, to an FMS control computer, there is little or no difference between one device being defined as a workstation or a small number of integrated devices being a workstation (or at least, dealing with the subtle differences is not a necessary part of this text). Typically, a workstation would have been operated by one person, if it had not been integrated into an FMS.

If one considers the hierarchy of computers within a typical factory, it is likely to be along the lines of that shown in figure 1.7.

Level	Function and computer type
Factory	Material requirements planning mainframe
Area/centre	Shop—floor supervisory minicomputer
Cell/system	FMS control system
Workstation	Ruggedised microcomputer or PLC
Process	Computer numerical control
Device	Sensors etc.

Figure 1.7   A typical factory computer hierarchy

Similarly, the types of 'mechanical' hardware that one might reasonably expect to find within a flexible manufacturing system may be categorised according to their function. For example, process equipment would be used to transform raw material into finished products. Transportation systems would be used to transfer parts between process workstations and either other process workstations or storage systems. Some examples of the equipment which would typically be found within these categories are listed below:

Process Workstations

- Machine-tools and their control systems (for example, machining centres, lathes, grinding machines etc.)
- Welding stations, perhaps with robots and other ancillary equipment
- Hole punching stations
- Assembly/disassembly stations (for example, parts being assembled, screw insertion, electronic component insertion into a printed circuit board etc.)
- Forging facilities
  . . . etc.

Transportation systems

- Conveyors (for example nylon flexible chain link, power and free, carry and free, slat etc.)
- Automated guided vehicle systems
- Fork lift trucks, pipes
  . . . etc.

Storage units

- Load/unload bays
- Automated storage and retrieval systems
- On-line buffer stations, storage racks
  . . . etc.

Similarly, the major elements of an FMS control system are usually threefold:

(1) The host computer system, whose task is to co-ordinate the activities of the various equipment (workstations) within the facility.
(2) A communications network, whose task is to link the host computer system to the workstations.
(3) Finally, control units responsible for the detailed activities of the workstations themselves — for example, a computer numerical control unit on a machine-tool or robot, or possibly a factory ruggedised microcomputer or programmable logic controller etc.

To confuse the situation slightly, it is currently often necessary to have a further device to interface the device controllers, say of the machine-tools, to the FMS communications network. These devices are generally called *interface units*, or for the purposes of this text, *workstation interface units* (WSIUs).

The fact that it is usually necessary to have a WSIU is indeed unfortunate and maybe in a few years' time it will in fact prove to be unnecessary. Currently the basic requirement for WSIUs results from the fact that most control units are not able to support true FMS communications — that is, to permit the transmission and receipt of part programmes, status information and control instructions. Many control manufacturers are trying to address this problem, as a part of their response to the MAP movement.

Fortunately, it turns out that having WSIUs does have its advantages, since such devices may also be used to provide, among other things, a standard operator interface to the host computer system, which might otherwise have been difficult and/or expensive to implement.

So these are the major control elements of flexible manufacturing systems. Figure 1.8 is a block diagram showing how these elements are combined within a typical installation.

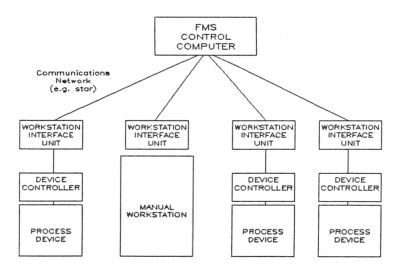

Figure 1.8   Typical FMS control elements

1.9   Classifying FMS*

While these are the major elements comprising an FMS, their presence within a manufacturing environment does not necessarily mean that they form an FMS.

*Much of the data for this section was made available by the kindness of Frost and Sullivan. The content of their market surveys has been interpreted in the light of the author's own experience.

The key questions to bear in mind when considering whether a facility is really an FMS or not (and if so how flexible it is) are essentially as follows:

(1) How flexible, or conversely how dedicated, are the workstations?
(2) How easy is it to change a workpiece's routeing through the system?
(3) How easy is it to alter the scheduled workload for the system?
(4) How easy is it to change the products that are being produced on the system?
(5) How easy is it to change the volume of the various products that are being produced by the system?

The answers to all these questions will be 'qualitative' rather than 'quantitative', since there are obviously varying degrees of flexibility in all manufacturing systems. Nevertheless, these questions may be used to distinguish the more sophisticated systems from the less sophisticated.

An additional question which may be used to help classify automation systems (not only FMS) is, how well is contingency management handled? If one workstation fails, does the whole system cease production? How easy is it to abandon manufacture of a batch of components which is already distributed throughout the FMS? How much assistance is provided to the operators when an item of equipment has failed and needs to be restarted? These questions are far from being as superficial as they might at first appear. In fact they are probably among the most important to be considered during the design phase of the FMS, as will be discussed later. Strangely, however, they sometimes appear not to be asked, to the eventual detriment of the system.

Certainly, the difficulty of classifying flexible manufacturing systems has made it more difficult to establish real trends in the development of this still relatively new approach to manufacturing. Asking the above questions helps to categorise systems, but not really to classify them. However, some general observations about the way in which flexible manufacturing systems have been installed to date are shown in figures 1.9 to 1.12.

It should be appreciated that much of this data is, of necessity, derived from the systems which are currently installed. Therefore when interpreting the data, the conclusions drawn reflect this installed base rather than current trends. For example, if one considers that the majority of flexible manufacturing systems currently installed are producing cubic components, some of the data becomes more understandable, if, unfortunately, less meaningful.

It is likely that during the next 10 years the uptake of FMS technology, in one form or another, will be substantial. The growth rate of system installation is forecast to be between 20 per cent and 50 per cent per year, both in Europe and in the USA. However, the number of systems installed is predicted to be slightly higher in the USA, with each system comprising slightly more process devices.

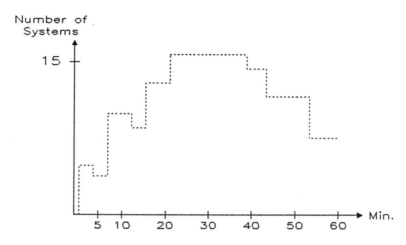

Figure 1.9   Typical FMS process cycle times

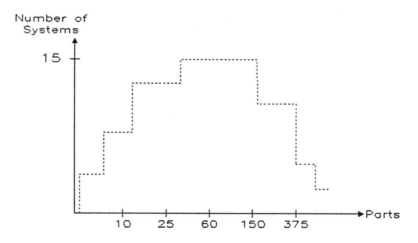

Figure 1.10   Typical FMS batch sizes

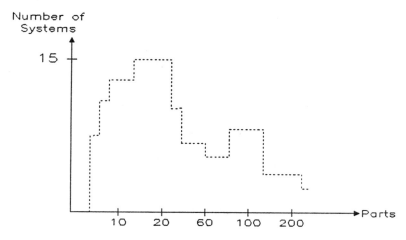

Figure 1.11   Typical number of different parts that an FMS produces

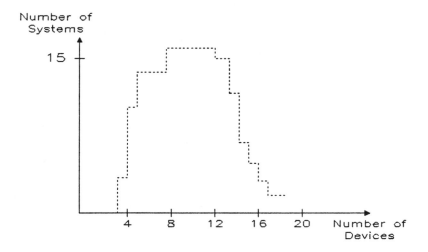

Figure 1.12   Typical number of process devices within an FMS

Also, the current bias of FMS towards aerospace and machinery industries' applications is likely to shift to automative and especially electronics, where flexible assembly systems are likely to become extremely important.

1.10   Why implement FMS?

The benefits that can be achieved by successfully implementing a flexible manufacturing system in the correct environment are quite substantial. Unfortunately, some of these advantages are rather more difficult to evaluate than others, and this tends to cause difficulties when one is trying to convince 'bottom-line oriented accountants' of the viability and necessity of such a system (a subject which is covered in some detail in chapter 10). However, a list of the advantages which might typically be expected to result, would certainly include the following:

(1) Improved capital/equipment utilisation.
(2) Reduced work in progress and set-up.
(3) Substantially reduced throughput times/lead times.
(4) Reduced inventory and smaller batches.
(5) Reduced manpower.
(6) Ability to accommodate design changes readily.
(7) Consistent quality
(8) Reduced risk as a result of specific product failure.
(9) Concise management control.
(10) Improved market image/credibility.
(11) Reduced floor space requirements.

Of course there are also some disadvantages, such as:

(1) Still a relatively new technology.
(2) Good design/implementation expertise is difficult to find.
(3) Systems are complex, necessitating lengthy operator and maintenance training.
(4) Systems are expensive.
(5) Systems take several years to implement.
(6) Currently systems require the writing of a significant amount of once-off software for the central computer system.
(7) Currently difficult to integrate devices from different manufacturers.
(8) Difficult to find capable/credible systems suppliers who are likely to be able to provide long-term support.

In time, many of these disadvantages will be addressed by the major automation system suppliers. This will be very good news for the devotees of

FMS (and there are an ever increasing number of these) since the advantages of FMS will continue to become more pronounced, while the effect of the disadvantages will be diminished.

2 Some Examples of Flexible Manufacturing

2.1 Introduction

This chapter describes a number of flexible manufacturing systems which have already been installed, some in detail and some purely in outline, the intention being to give the reader an impression of the size, complexity and variety which is apparent in these sophisticated production facilities.

2.2 The SCAMP flexible manufacturing system

During January 1983 a flexible manufacturing system (FMS) started producing parts in Colchester, England. Codenamed SCAMP, standing for the Six Hundred Group's Computer Aided Manufacturing Project, it marked the start of an era within Europe of the design and implementation of sophisticated, truly flexible manufacturing systems. The system was capable of producing a wide variety of relatively simple turned components, typical of those used throughout engineering industries worldwide. The system was developed by SCAMP Systems Ltd, a wholly owned subsidiary of The 600 Group PLC.

Interestingly, the project started following a UK Government initiative to attempt to interest manufacturers, particularly within the machine-tool industry, in the rebirth of FMS technology. Sir Jack Wellings of The 600 Group took up this challenge. After a feasibility study had been carried out, which included a world tour of installed systems, a proposal was submitted to the UK Government for funding to assist with the development of a world class FMS. It was for a flexible turning system, initially to be capable of producing about 40 different parts, but with an ultimate capability of several hundred different parts.

Funding support was made available and, between 1979 and 1982, the SCAMP system was designed, installed and commissioned on its present site in Colchester, near the well-known Colchester Lathes plant (for which, incidentally, many components are now being made by the FMS). The total cost of the system was some $4.5 million.

SCAMP as finally implemented comprises 9 machine-tools, 8 robots, a sophisticated 'carry and free' conveyor system and a complex hierarchical computer system (see figures 2.1 and 2.2). The main objective of the SCAMP system was to prove that it is practical and cost-effective to manufacture turned components in small batches, with a total throughput time of days rather than months. In fact, throughput time was reduced from three months (in the efficient traditional plant where the parts were still being produced) to only three days.

SCAMP is actually two flexible turning systems which share their first operation machines, these being two 2-axis CNC lathes and two 5-axis turning centres. After visiting these machines, components follow one of two processing routes, depending on whether the parts are 'disks' (that is, gears, pulleys and bearing covers) or 'shafts' (see figure 2.3). The first leads to a gear shaping, toothrounding and deburring bay followed by a horizontal broaching bay, for 'disks'. The alternative is a cylindrical grinding bay and a spline milling bay, for 'shafts'. Wherever possible, work is performed by machines in the order in which they are physically located in the system. This is why all the turning bays are located at one end of the system, with the support bays following in the direction of travel of the main conveyor system. However, if

Figure 2.1   An overall view of the SCAMP FMS
(courtesy of SCAMP Systems Ltd)

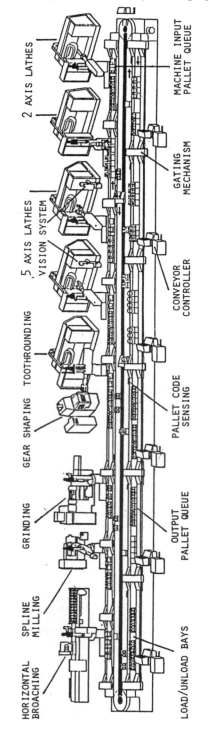

Figure 2.2   An overall schematic of the SCAMP FMS (courtesy of SCAMP Systems Ltd)

(a)

(b)

*Figure 2.3   (a) The SCAMP rotational components
(courtesy of SCAMP Systems Ltd).
(b) For comparison, a typical group of cubic components
(courtesy of Deckel)*

necessary, for a particular part, a completely random process route can be accommodated, albeit at the imposition of some additional conveying time.

At the heart of SCAMP's flexibility is the conveyor system. This system, while mechanically relatively straightforward, is highly sophisticated in control terms, in order to meet the requirement for considerable routeing flexibility.

The conveyor system consists of six separate conveyors, four parallel chain conveyors and two turntables. The chain conveyors comprise two parallel chains linked by a number of free-running rollers, positioned approximately four inches apart. These conveyors are in motion whenever the system is in operation. To ensure that friction from the guide-rails does not prevent objects from being moved by the conveyor, the rollers are spring-loaded so that the forward force on the pallets being transported is increased slightly. But if the object is intentionally restrained, only a small load is exerted by the conveyor as it continues to move beneath the pallet.

Each conveyor, however, moves in the opposite direction to its neighbour(s). The two outer conveyors are subdivided by gating mechanisms to form a number of individual sidings, six load/unload sidings and eight machine buffer sidings. The gating mechanisms are responsible for transferring the pallets between the two types of sidings and the other main conveyors.

The final two conveyors are in fact turntables. These are placed at each end of the two central chain conveyors. Together, these four central conveyors (that is, excluding the two types of sidings) combine to form a complete recirculation loop around which pallets are able to travel until they are required to be gated into another part of the system.

All the parts are transported within SCAMP on pallets that have been designed to ensure that an absolute minimum of tooling (that is, one fixture) is needed to accommodate all the parts capable of being manufactured. If the components are 'disks', they are simply placed on top of the pallets. The upper steel plate of the pallet is coated with a mild abrasive to counteract any tendency of the parts to slip. Depending on the size and stability of the parts, pallet capacity ranges from one to sixteen parts. This includes the vertical stacking of parts where possible. If the components are 'shafts', then a simple steel fabrication is secured to the top of the pallet which allows up to nine shafts to be held vertically in 'pots' formed by three equispaced spring-steel strips. A horizontal base plate can be slid into one of three vertical positions to facilitate the accommodation of shafts of different lengths.

The pallets are all uniquely coded during manufacture. This is achieved by fixing a number of small steel tabs along a strip in the side of the pallets. The computer system is therefore able to relate a particular set of parts and its stage of manufacture with a specific pallet. In fact, the design is such that, in normal circumstances, it is not even necessary for an operator to enter a pallet number when carrying out the manual load operation. The main computer system is able to correlate the load instructions it issued with the number of the pallet which appears at the next pallet reading station after the gating of the loaded pallet on to the main conveyor has occurred.

The manual load/unload bays (see figure 2.4) are the normal operator interface to the system. The operators receive instructions as to what to load via a standard VDU. These instructions comprise text and pictorial representations of where parts should be placed and perhaps stacked on the pallet. To respond to these instructions and generally to communicate with the main computer system, the operator uses a ruggedised terminal called a 'conveyor controller'. Conveyor controllers are essentially microcomputers that are used to control a particular section of the conveyor. There are fourteen of these units located around the conveyor. This subdivision and distribution of the conveyor control task substantially reduces the burden of routeing the pallets throughout the conveyor, on the main computer.

While the six load/unload bays are all of equal capacity, the eight machine buffer sidings vary in length. The individual capacities of these were decided

Figure 2.4   A SCAMP load/unload bay
(courtesy of SCAMP Systems Ltd)

on the basis of a number of computer simulations. It is worthwhile noting that a significant amount of simulation was carried out by a number of different teams to assist with the design of SCAMP. This work was inexpensive relative to the total investment involved, and yet proved to be invaluable.

Since the two outer conveyors move in opposite directions, pallets destined to visit a machine must travel past the machine on the main conveyor until they reach the gating mechanism which is able to transfer them into the appropriate machine buffer siding. Once in the siding, the pallets, by the action of the motion of the conveyor, are automatically added to the queue of pallets waiting for the machine. As pallets of machined parts are returned to the main conveyor by the gating mechanism at the other end of the siding, so the pallets advance towards the machine. At the end of the machine end of the siding, three air-operated latches hold pallets in specific positions (see figure 2.5). The last position is for the pallet on to which the robot will load finished parts. The central position is where the pallet from which the robot unloads raw material rests, and the final position marks the beginning of the queue of outstanding work for the machine. These latches separate the pallets by about six inches to ensure that when the versatile, but relatively large, robot gripper moves a part on one pallet, it does not sweep off parts from a neighbouring pallet. Also, since parts are stacked on pallets, it is essential

Figure 2.5   A SCAMP machine buffer siding
(courtesy of SCAMP Systems Ltd)

that components are loaded and unloaded from different pallets. If this was not the case, either parts would have to be temporarily stored within the bay, or the pallets could not be loaded to capacity.

At the same time, to help speed up the loading and unloading of the machines, there is a buffer for two parts within each bay, close to the machine. One of these is for a new piece of raw material and the other for a finished machined component. This saves the robot from having to return to the pallets during a machine load/unload cycle. Instead the robot only needs to travel between the buffer stores and the machines. This minimises the idle time imposed on the machine-tool while it waits for a new part to be loaded.

Having separate load and unload pallets means that while material passes through a bay, the majority of the parts will be transferred from one pallet (unload) to the pallet in front (load). However, this does not apply to the two parts in the buffer stores. This is because the pallets are 'moved up' when the unload pallet is empty — that is, the load pallet is returned to the main conveyor and the now empty unload pallet is moved up to become the new load pallet. All this occurs while the two last parts from the old unload pallet are still in the buffer stores, thus posing a complex tracking problem for the main computer. Yet a further consequence of this approach is that an empty pallet is needed to lead the pallet train into the bay; otherwise there is nowhere to put the components which are machined first.

Inspection of components is carried out manually at the inspection station located within each of the machining bays. An operator is able to request the next machined part for inspection. Alternatively, during the input to the computer of the parts' processing requirements, the computer can be instructed to ensure that, say, one in every twenty parts is inspected.

Although work was undertaken to equip all the lathes with in-process gauging based on probes, this did not initially prove to be successful. Only the cylindrical grinding machine was fitted with full in-process gauging. This proved most effective, significantly reducing scrap and hence the need for manual inspections in this bay.

The two turning centres were equipped with a vision system. This was used to orientate asymmetric components prior to their being placed in the machine (see figure 2.6).

The robots used to transfer the components between the pallets and the machines are all of the cartesian type, manufactured by Fanuc. These were interfaced to the main computer by a factory ruggedised microcomputer similar to that of the conveyor controllers. This unit not only facilitated communications between the robot and the computer but also acted as an operator interface to the computer, dedicated to a particular machine bay. The terminal may be used by the operator to perform a variety of functions. For example, self-diagnostics, starting the bay after a set-up, inputting the results of or requesting an inspection, not to mention inputting data used within the contingency management functions, such as recovering a bay after an equipment failure etc.

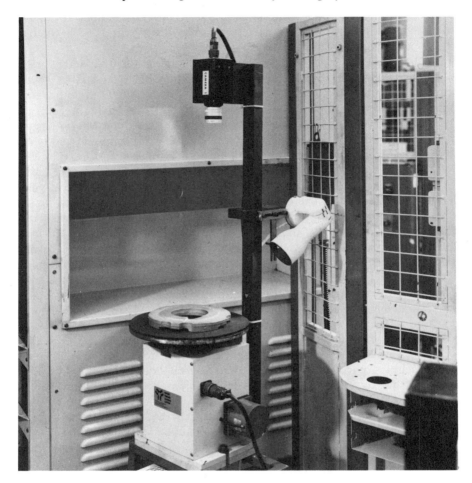

Figure 2.6   The vision system used within SCAMP
(courtesy of SCAMP Systems Ltd)

All the major equipment within the system communicated with the host computer system via a specially designed DNC network. This was implemented in a 'star' configuration (see chapter 9), the centre of which is a switchover unit which directs the communications to one of the two computers capable of controlling the system. The development of the communications protocol and also the watchdog timer (which formed the basis of the hot standby facility, see chapter 8) necessitated a significant amount of effort. However, the result was an impressively effective system (refer to figure 2.7). If the 'live' processor failed, the 'back-up' could continue to run the system after only a few seconds. This was an expensive option but essential if true twenty-four hour operation is required.

*Figure 2.7   The author operating the SCAMP computer system
(courtesy of SCAMP Systems Ltd)*

The hierarchical computer system comprises two PDP 11/34 equivalent computers, the communications switchover unit and sixteen factory ruggedised microcomputer terminals. To interact with the main computer system, operators use a fairly typical menu-driven system on black and white VDUs. Colour graphics displays in the main control room are used for two tasks. First, for a mimic display, where a representation of the real-time status of the equipment within the system can be obtained. Second, for the work scheduling system. This uses the colour display to show a 'Gantt chart' of the work scheduled for the system. Unique colour codes are used to represent individual part types, and batches of these parts are then represented by blocks of this colour. The length of the block denotes the time required for the particular operation, each machine-tool's schedule being represented by a horizontal line on the screen. Some intelligence was built into this system to facilitate the operator's task of scheduling work into the system. Ultimately, however, the system still relies on the skill of the operator to provide any optimising of the workload for the system.

A detailed discussion of the capabilities and complexities of the software needed to control the SCAMP system would merit a book in itself. While this might well be interesting, it is not the purpose of this book. It is worth noting, however, that the software required to cater for contingency management (for example, start-up, close-down, abandon batch, recovery and component tracing after inspection etc.) accounted for a substantial proportion of the software effort, a factor which is often ignored by FMS designers.

In conclusion, the SCAMP system proved to many how viable FMS technology had become. It remains an impressive system and an outstanding achievement, especially when one considers that the system was implemented by a very small full-time project team. The value of the SCAMP system is that it has clearly demonstrated to the large number of companies which operate machine shops, however small, that FMS is both a practical and an economic reality.

2.3 Vought Aero Products Division FMS

This system, located in Dallas, Texas, is both new and sophisticated, having only recently been commissioned. The cost of the system, which manufactures complex cubic components, was approximately $10 million. This covered all the equipment, the computer hardware and software, and the installation. This sum also includes the cost of preparing an existing building to house the FMS. The work was carried out by Vought Aero Products (a division of the LTV company) that currently employs about 15 000 people, 11 000 of whom are located at the main Jefferson Avenue plant, where the FMS is sited.

The system was supplied as the direct result of a full turn-key contract being placed with Cincinnati Milacron. This company had the responsibility of subcontracting those elements of the system which it was not able to supply itself. Vought furnished the support equipment, such as part storage racks, the tooling ASRS etc.

Vought initiated the project in order to rationalise and upgrade its manufacturing capabilities. This process was started during the early 1970s. Since that time, Vought has become one of the leading US aerospace subcontractors, as opposed to being a prime supplier of aircraft. This change in philosophy naturally required an equally substantial change in manufacturing practice. In particular, a need for higher efficiency and productivity was evident. This represented both a challenge and an ideal opportunity to apply sophisticated automation equipment.

The eventual aim of Vought is very bold: it is to have the entire plant as one fully computer integrated manufacturing facility. Obviously, this is a long-term goal, since the problems of integration are substantial and unlikely to be solved overnight. Nevertheless, Vought is to be commended on at least having its ultimate goal in mind from the outset.

Vought is responsible for the manufacture of about 2000 machined components for the B-1B fuselage. Of these, approximately 540 were selected for manufacture within the FMS. This resulted in the preparation of in excess of 2100 part programs. By the time the system was commissioned, these programs had all been written, though not all had been tested. This was, of course, a sizeable task in its own right.

In summary, the system comprises eight machining centres, two co-ordinate measuring machines, a washing machine, an AGV system, two manual loading and unloading carousels and a tool storage ASRS. Much of the FMS was designed with the aid of computer simulation. This was carried out by Vought using an IBM PC-based simulation package with animated colour graphics.

On average, the components have a cycle time on a machine of approximately 20 minutes. However, the minimum cycle time could be as low as 5 minutes with the maximum as high as 60 minutes. On average, each part visits the FMS 2.3 times before it is complete, though this does not include a visit to the automatic gauging machine which is visited by all parts as their last operation.

It was approximately 18 months after the order was placed before the first components were manufactured by the system. This was during mid 1984. Full operation of the system was scheduled to start during July 1984. Unfortunately, owing to control software delays, the opening of the fully enhanced FMS did not occur until late 1985. However, a subset of the FMS's capabilities was used for production from the originally planned starting date.

The machining section of the FMS comprises eight identical Cincinnati Milacron, 4-axis simultaneous contouring, single-spindle (40–5000 r.p.m) machining centres. These are responsible for carrying out all the machining, boring, drilling, reaming, tapping and profiling required to complete the parts. The working envelope of the machines is 36 inches × 36 inches × 32 inches. All the machines are equipped with automatic tool changers with a capacity of 90 tools, these being located in two carousels each of 45 tools. The machines are also equipped with an automatic tool gauging system.

Each machine-tool control system is equipped with its own communications link to the main computer system. This is used to provide real-time status information as well as to support the transfer of machine-tool part programs. In addition to operating the machine-tool, the controller, in conjunction with a programmable logic controller, is responsible for the operation of the pallet transfer mechanism. This is used to transfer the components, which are mounted on pallets, between the machine-tool and the automated guided vehicle system. The pallet transfer mechanism has a capacity of five pallets: two for input, two for output and one in the machine.

Tools are manually pre-set, being stored in a dedicated mini-load ASRS. Picking requests for tool sets are sent to a computer terminal located close to the ASRS. The operator who receives the request extracts the appropriate tools and loads them on to a tray. The tray is then placed on a cart, which is moved manually into the FMS. Currently, loading of tools into the machines carousels necessitates the interrupting of the cutting cycle. However, since this work is usually carried out only during a set-up operation, when the machine is idle anyway, it does not impose a large penalty on machine availability. Nevertheless, an enhancement is planned (the first stage of which is already being installed) whereby tools will be delivered automatically to the machines. Furthermore, the used tools will automatically be unloaded and replaced with those required for the next parts.

The material handling system which serves all the machine-tools and ancillary equipment is an AGV system supplied by Eaton–Kenway. It includes four 'robocarriers'. In fact, the simulations carried out concluded that three such carriers were needed, and hence four were purchased to provide adequate coverage for breakdown.

The loading and unloading of parts to and from pallets are carried out at two manual loading stations (see figure 2.8). These are carousels with a capacity to hold up to ten pallets. Load and unload instructions are sent to factory ruggedised terminals located close to these stations. Additional information, for example, about the fixturing and orientation of the parts to be loaded is displayed on a microcomputer-based colour graphics terminal, also near the loading stations. This Computer Aided Graphic Instruction Network was added later by Vought.

The fixturing of components is relatively flexible. Seven types of fixtures are used. These are assembled to meet the specific requirements of the pallet from modular sets, according to the instructions sent to the operator. Up to sixteen different components could be present on a pallet at any one time.

A centralised coolant distribution and swarf removal system is incorporated into the FMS. Two underfloor channels are used continuously, one for aluminium and one for steel. The machine-tool controllers are able to direct swarf into the appropriate channel through instructions included within the part programs. Synthetic coolant is used throughout. Initially, problems were encountered with both the containing and removal of swarf. Some redesign of the machine-tool guards was needed to overcome these problems. After a part has been machined, it is always sent to the washing machine to remove any remaining swarf and prepare for the inspection operation.

The main burden of quality control within the FMS is the responsibility of two DEA Delta 3404 gantry-type co-ordinate measuring machines. These machines are fully computer-controlled and operate within the normal FMS environment (which in this case is impressively clean). However, owing in particular to the proximity of these machines to the machining centres, floating foundations are used to eliminate the effects of vibration.

If a part is measured and found to be out of tolerance, a complex corrective procedure is invoked. First, the part suspected to be out of tolerance is automatically routed to the second co-ordinate measuring machine by the control computer. If the part is rejected after a second inspection on this machine, the control computer instigates a number of actions. A dynamic 'hold' is placed on the NC part program used to manufacture the part originally (this reduces the probability of creating further incorrect parts). A hold is also placed on the machine-tool used to create the part. Finally, messages are sent

Figure 2.8 An overview of the Vought FMS (courtesy of Vought Aero Products Division)

to the personnel responsible for putting together the cutter kits which were used to manufacture the part.

After this, a calibration cube is sent to the relevant machine-tool and, by using a probe mounted in the spindle, Vought check the accuracy of the machine. The cutter kit is extracted from the carousel and checked for any faults. Meanwhile, the NC program (which would need to have been certified prior to having been used for production) remains on hold until the investigation is completed. Once the problem has been identified and rectified, the control computer releases the various hold states, permitting the scheduling of further work and the continuation of machining. A manual material review station is used to investigate whether a damaged part can be salvaged, or must be scrapped.

The computer system hardware used to control the FMS comprises a Digital Equipment Corporation (DEC) PDP 11/44 with an 11/70 available if needed as a 'cold' standby. A link exists between this system and Vought's business systems mainframe, on which resides the overall plant work schedule.

Five people are needed to operate the FMS during any of the five days per week and three shifts per day that it is operated: one person to patrol the machines, one located within the control room, one for handling the tooling system, and the last two being responsible for the loading and unloading of parts. If a manual inspection is required, an additional operator can be made available.

The implementation of the facility was justified on the basis of savings in excess of $20 million for the US Air Force in the cost of manufacturing the required components for some one hundred B1 bombers. These parts are all expected to be produced over a two year period.

Without doubt, this system is extremely impressive. Vought's claims of this being a "world class system which represents the leading edge of technology" are probably justified. Like many similarly advanced aerospace manufacturing companies, Vought has a number of sophisticated islands of automation. Joining them together as a part of a true CIM system will not be easy.

2.4 FMS within General Electric (USA)

General Electric USA (GE) is one of the most diverse, if not the most diverse, manufacturing company in the world. Based in the USA, but with substantial interests in Europe and elsewhere, GE is large by almost any standards:

- 1984 turnover, $28 billion.
- 350 manufacturing plants.
- 340 000 employees worldwide.
- 1984 automation investment, $2.5 billion ($8 billion between 1980 and 1984).

- 1984 investment in Research and Development, $2.3 billion.

However, not only is GE a large diverse manufacturer, it is also one of the major world manufacturers of automation equipment and systems.

With so much being invested in automation equipment, it is hardly surprising that a substantial proportion is devoted to the development of flexible manufacturing systems. GE's size and product diversity (consumer electronics, steam turbines, household appliances, aircraft engines etc.) help to make the company an interesting case study. If GE was not able to automate effectively, not only GE's automation business would suffer but also its other businesses. The common factors to the following automation success stories are clear: management commitment, long-term strategies and, above all, enthusiastic and competent people.

Few would deny that the decline of many industrial organisations, and possibly even nations, has been stimulated (if not caused) by their failure to develop and use effective automation strategies. General Electric was determined not to allow itself to fall into this category, but to take advantage of the possibilities offered by the substantial advances being made in automation technology.

For example, when in 1979 a market survey showed that there would be a substantial increase in the USA locomotive market during the mid 1980s, the management of GE's Transportation Department in Erie, Pennsylvania, took the decision to embark upon an ambitious automation plan in order to be able to secure a substantial proportion of the forecast market.

Almost immediately, it was concluded that a traditional 'piecemeal' approach to the automation would be unlikely to yield the required results. Instead a 'total systems approach' was needed. Hence a multi-discipline, inter-department team was formed to develop a long-term automation strategy. The ultimate objective was to improve productivity, product design and quality.

Essentially, the scope of the project was to modernise ten facilities within two separate plants (Erie and Grove City). The total cost of the program was to be in the region of $300 million. This was to be invested in three phases, spread over some seven years. Not surprisingly, both products and processes had to be redesigned to enable the program to be a success, but this having been done, the results were startling:

- 33 per cent increase in production capacity.
- 20 per cent material savings.
- Substantial reduction in manpower requirements.
- 33 per cent reduction in assembly line costs.
- 20 per cent increase in inventory turnover.
- Substantially improved product quality.

Within the Erie Plant, frames for GE's Diesel Electric Motors are now being machined within a sophisticated FMS; these are large fabricated parts, approximately 36 inches × 36 inches × 54 inches (see figure 2.9). The system comprises nine machining centres all supplied by Giddings & Lewis. These are:

- *Three MC 70 machining centres*
  These have 7-inch spindles and perform the rough boring and milling operations. The machines are equipped with automatic tool-changers each with a capacity of 100 tools.
- *One MC 60 machining centre*
  This has a 6-inch spindle and carries out the finish boring and milling operations, again with an automatic tool changer of 100 tool capacity.
- *Three 15 HS machining centres*
  These machines, which are again equipped with tool storage and automatic tool changers, are responsible for the drilling and tapping operations.
- *Two special vertical milling machines*
  These machines are unique to the system in that they share a static tool store, which is replenished manually, but is also served by a robot which passes the tools to the machines. The machines are also equipped with an index table and compound slide assembly which provide access to the inside of the components, notably the inside pole surfaces and the brush holder pads.

Figure 2.9   A motor frame about to be machined in GE's Erie FMS
(courtesy of General Electric (USA))

One of the key factors in the selection of the machines was that many of the operations could be interchanged. This considerably simplifies the scheduling task of the FMS control computer system, and facilitates high machine-tool utilisation. It also provides additional system resilience to individual machine-tool failures.

The machines and the manual load/unload datuming station are all served by a single rail-guided cart, transfer mechanism.

The whole system is controlled by two DEC PDP 11/44 computers on cold standby (that is, an operator is required to start the standby computer in the event of the live computer failing).

The raw materials (in the form of welded fabrications) are delivered to the FMS by fork-lift truck. From here they are transferred to the load/unload station, which is also the location of the automatic component datum device. This transfer is a relatively delicate task, not only because the parts are large and heavy, but also because they are valuable (about $2500 at this stage). Once on the datum device, an expanding hydraulic mandrel is lowered into the part (see figure 2.10). This is used to locate the component accurately on the pallet. Once the part is correctly located, the operator clamps the part in position. It is then automatically fed into the system for the first of three series of machining operations. During each phase the part visits several machines, eventually being returned to the load/unload station for an operator to reorientate the part ready for the next phase of the machining processes. It has been said that this automatic qualification of parts prior to the machining has been instrumental in the success of the system. Ensuring that the workpiece has been located optimally from the outset, together with good control of the machining process, has resulted in the virtual elimination of post-machining rectification.

The system was designed to produce a family of eight parts. This has already been increased to fifteen. Despite this, only three sets of modular fixtures are needed. The cycle times on the machines vary between 2 and 120 minutes. Production tooling is monitored by the control computer and automatically transferred into the machines from buffer stores, though an operator is required to replenish these stores manually.

The lead times for these parts has been reduced from sixteen days to approximately sixteen hours. In addition to this, there have been substantial component quality improvements which have exerted a significant beneficial influence on many other stages of the locomotive production cycle.

The number of machines required to manufacture these parts was substantially reduced. It decreased from 29 to 9. Similarly, the number of operators required was reduced from 78 to 13 (5 being needed to run the first shift, and 4 on each of the other two shifts). There was also a 40 per cent saving in factory floor-space requirements and considerable inventory reductions. Currently, quality control is essentially manual, but there are plans to automate this soon.

Interestingly, stand-alone CNC and transfer line facilities were both rejected in favour of FMS, essentially on the grounds of lack of flexibility and productivity. So much confidence exists in the viability of this system that space has been left for additional machines. Even the rail-guided transfer mechanism is capable of accommodating another cart.

Approximately $16 million was invested in this system over a period of about four years. The project was started in 1979 with the system being installed during mid 1982. It has been running in a production environment since mid 1983. Quite rightly, General Electric is very proud of this system, one of its earliest ventures into FMS (see figure 2.11).

About 100 miles south of Erie is the new diesel engine plant at Grove City. GE's goal with this plant is to make it into the most automated diesel engine production facility in the world. It will be a true 'paperless' factory, built using the experience gained with Material Requirements Planning (MRP) systems at the Erie facility together with advanced machining facilities. It is forecast that productivity will be increased by a factor of nearly 2.5, while throughput time will be reduced to approximately 15 per cent of previous experience. Accordingly, inventory is expected to reduce by half because of the combination of new technology and the 'Just-in-Time' (JIT) approach to production control. Product quality is expected to improve considerably owing to the improved processing equipment and integrated computer quality support systems.

While GE's Erie and Grove City facilities are concerned with physically large parts and relatively low volumes (producing some 5500 parts annually), at the mass production end of the FMS spectrum is the Major Appliance facility at Louisville in Kentucky. At this plant dishwashers and other domestic appliances are manufactured.

It was found that the market share of this business was declining rapidly owing to problems of quality and reliability. So a plan was devised to modernise and automate the existing 30-year-old facilities. $38 million was invested over the four years between 1979 and 1983. A very similar approach was taken to that adopted in Erie, despite the widely differing production environments — evidence that the FMS philosophy is largely independent of product type and volume requirements. The results are equally impressive as those obtained within Erie:

- 10 per cent reduction in the cost of labour, materials and overheads.
- 33 per cent reduction in cycle time.
- 40 per cent reduction in scrap and rework.
- 25 per cent increase in productivity.
- 78 per cent fewer parts.
- Significant reductions in inventory and handling charges.

Certainly the facility could never be regarded as having the same degree of flexibility as say the Erie FMS, but it is capable of producing substantially higher volumes. The flexibility of the system at Louisville lies in the ability to accommodate design changes readily and switch production between differ-

*Figure 2.10   A part being aligned at the datum station
(courtesy of General Electric (USA))*

Figure 2.11   An overview of GE's Erie FMS   (courtesy of General Electric (USA))

ent models. This has been made possible by the careful combination of a variety of technologies and techniques; for example, a factory management system, interactive graphics, closed-loop machining, robotics and automated materials handling etc.

One of the most interesting facts about the Louisville facility is that while the tangible benefits, predicted and obtained, from the investment have been substantial, the intangible benefits, which would have been virtually impossible to quantify at the outset of the program, have proved to be even more important. Before the automation program, GE's dishwashers were encountering a severe reduction in their market share; now, the demand is in danger of increasing beyond the system's capacity to supply. The ability to be able to guarantee both quality and delivery dates has proved so valuable to the marketplace that this factor alone, which was not forecast within the project evaluation, would have justified the investment.

To conclude this section, two of GE's best known automation programs have been discussed. They have resulted in significantly different types of FMS being installed. Both incorporated a 'total systems, long-term approach', hardware, software and people. These are all key requirements for both success and survival.

There are many more sophisticated flexible manufacturing systems within GE, some installed and others under construction, particularly within the Aircraft Engines Business Group. Unfortunately space is not available within this book for a description of them all.

2.5 FMS at Fanuc

The Fanuc company of Japan is one of the automation success stories of our time. Not only for the highly successful range of automation products which has been able to capture a substantial proportion of the automation market, but also because of the highly sophisticated automated production facilities which are used to manufacture its products.

Fanuc was incorporated in 1972, with 620 employees. Sales in that year amounted to $28 million. In 1985 sales had risen to $660 million, while the number employed was only 1500. This is an outstanding achievement, and is largely due to the commitment of the people, the long-term approach to the business, and the successful application of advanced technology within both products and production.

There are five Fanuc factories (one for each of the major product lines) where FMS, in particular, has been applied to considerable advantage: the motor factory, the robot factory, the wire-cut machine factory, the CNC factory and finally the injection moulding machine factory. While all these facilities are worthy of a detailed discussion, this text will only be devoted to the motor and robot factories. Only a brief overview of the other facilities will be included. The main aim of this section of the chapter is to show that FMS need not be restricted to an individual cell or system, but can be applied throughout an entire factory.

Fanuc's motor factory is located within a special-purposed two-floor building. On the lower floor is the machining area, comprising 60 machining cells with 54 robots. On the second floor, assembly operations are carried out. Again the manufacturing process has been subdivided into cells. These assembly cells, which incorporate one or more robots, are arranged in a line. The cell robot picks up parts from a pallet which will have been delivered by the AGV system, and carries out an assembly operation. When the work is complete, the part is made available for collection, and so the cycle continues. The two floors are linked by an automatic warehouse which is used as a temporary store for machined components (see figures 2.12 and 2.13).

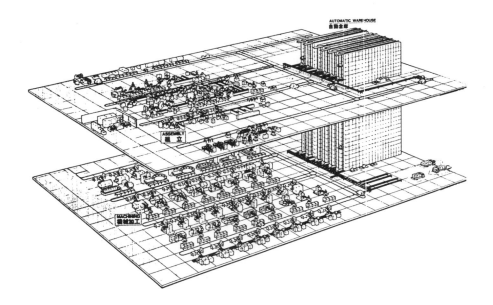

Figure 2.12   A schematic layout of Fanuc's motor factory
(courtesy of Fanuc Ltd)

Only 60 employees are needed to operate the entire facility. These, together with 120 robots, are able to manufacture 50 different types of motors at the rate of 15 000 per month.

In the robot factory 80 operators produce parts for some 550 robots, wire-cut machines and CNCs. In addition, 350 complete robots are manufactured each month.

The special-purpose robot factory is on one floor (see figures 2.14 and 2.15). In the machining section, 50 cells are automatically loaded and unloaded by robot. During the third shift, the facility is operated unmanned.

Figure 2.13   Inside Fanuc's motor factory
(courtesy of Fanuc Ltd)

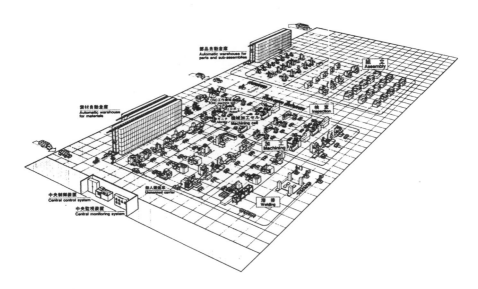

Figure 2.14   A schematic layout of Fanuc's robot factory
(courtesy of Fanuc Ltd)

Figure 2.15   Inside Fanuc's robot factory (courtesy of Fanuc Ltd)

Work transfer throughout is by Automated Guided Vehicle (AGV), with two automatic warehouses being used for buffering the supply of raw materials and subassemblies respectively.

In the wire-cut machine factory, only 30 operators are needed to manufacture in excess of 100 machines per month. While in the CNC factory, testing of integrated circuits and other components, parts insertion, printed circuit board (PCB) inspection, assembly and final testing are all carried out automatically. The facility has a monthly capacity of 8000 machine-tool CNCs, 1000 robot CNCs, and 1000 part program preparation systems. The injection molding machine factory is able to produce 300 machines each month. Both facilities are special purpose, and are run by only a few operators.

Fanuc's achievements are truly remarkable. Admittedly, the problems of substantial existing facilities having to be upgraded were not encountered, since all Fanuc's plants are 'green field' sites. Nevertheless, the degree of automation and unmanned operation is formidable. Probably the keys to its success are twofold: attention to detail during the implementation phases and long-term management commitment from the outset.

2.6 Examples of sophisticated automation systems

The final section of this chapter will be devoted to an overview of two machine-tool manufacturers' approach to advanced automation systems: OKUMA of Japan and Werner Kolb of West Germany. While it is debatable whether any of these systems possesses the features necessary to be classified as a true FMS (such as sophisticated scheduling and error-recovery capabilities etc.), they are nevertheless good examples of the type of highly automated subsystems which frequently form a starting point for the development of a true FMS.

The Okuma Machinery company is renowned for producing high-quality turning machines. Like virtually all machine-tool manufacturers, Okuma has expanded into the systems market in an effort to meet its customers' requirements for complete solutions to manufacturing problems.

Figures 2.16 and 2.17 show respectively a schematic layout and an overview of a fairly typical automated cell. While not being sufficiently complex to justify being called an FMC, it does illustrate a typical component of an FMC.

This particular system consists of two Okuma centres, a gauging station, a load and an unload station and a chamfering and cleaning device. All are served by an ASEA robot.

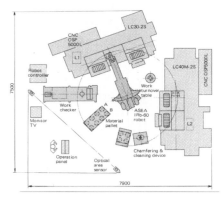

Figure 2.16   Layout of a small cell by Okuma
(courtesy of Okuma Machinery Ltd)

Figure 2.17   Overview of a small cell by Okuma
(courtesy of Okuma Machinery Ltd)

A slightly more complex system is shown in figures 2.18 and 2.19. In this system there are four machining centres and two lathes. Work transfer is handled by two Cincinnati Milacron robots and two short stretches of conveyor.

Werner Kolb has an established reputation for the manufacture of sophisticated flexible machining systems. Figure 2.20 shows a schematic layout of such a two-workstation FMC.

This type of system has been designed to accommodate pallets of either 25 inches × 31 inches or 31 inches × 39 inches. The main work transport system is a rail-guided cart mechanism, although a wire-guided system could also be used. The tooling system is particularly sophisticated: it includes a numerically controlled double-portal handling system, which is responsible for the transfer of tools between the machine-tool magazines and the central tool storage system. A double gripper capable of pivoting on two axes is used to help speed-up the tool change cycle.

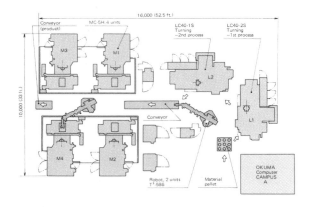

Figure 2.18   Layout of an FMC by Okuma
(courtesy of Okuma Machinery Ltd)

Figure 2.19   Overview of an FMC by Okuma (courtesy of Okuma Machinery Ltd)

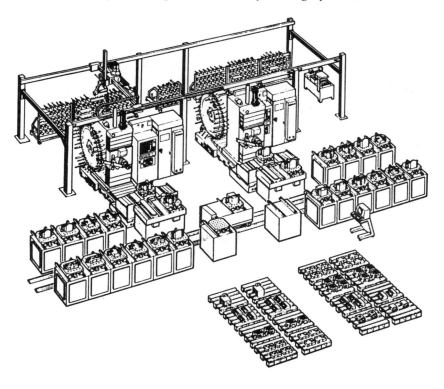

Figure 2.20   Schematic of a double workstation FMC by Werner
(courtesy of Werner Kolb)

A key consideration in this design is that it is modular. Further units can quite easily be added to extend the capabilities and capacity of the system. Figure 2.21 shows a schematic layout of such a system comprising four machining centres, tool and pallet storages, and a rail-guided transfer mechanism.

2.7   Concluding comments

The above examples of successful flexible manufacturing systems and sub-systems exhibit some common features that are worthy of note.

In general, they have not been installed on the basis of strictly meeting specific investment and return criteria, though substantial returns were achieved. In addition, they proved to be strong marketing tools, significantly leveraging the market's perception of the investing manufacturer. In particu-lar, they were undertaken with substantial investment of both time and

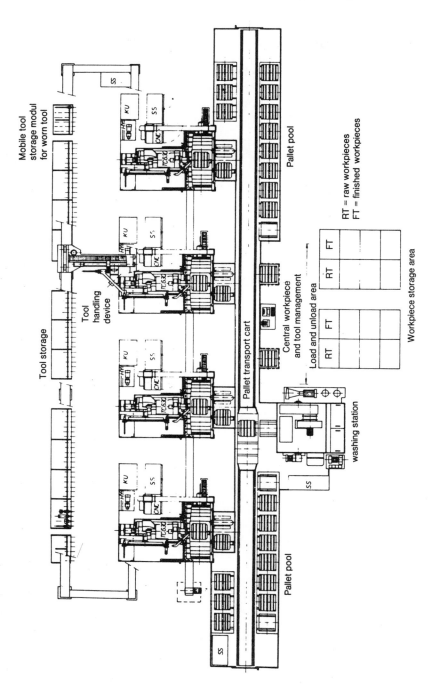

Figure 2.21   Layout of a four-workstation FMC by Werner (courtesy of Werner Kolb)

resources, in both detailed planning and computer simulation. They did not rely on an esoteric approach to hardware or software design and application, but nevertheless achieve a harmonious and elegant design solution to the integration of mechanical hardware and computer systems.

The Vought and SCAMP examples in particular are of immense significance to the large number of component manufacturing companies. Both are clear and competent demonstrations of a new era in efficient and fully productive manufacturing.

3 Project Structure and Management

3.1 Introduction

The success of an FMS is dependent to a large extent on the depth of consideration given to three issues:

(1) Whether the FMS is to manufacture the correct parts in the correct volumes.
(2) Whether the design is correct.
(3) Whether the project is managed correctly.

The last of these factors also has a significant impact on the first two. If the project's structure and management is not implemented well, then the FMS will fail.

One fact, which is perhaps slightly surprising, is that much of the overall implementation of any sophisticated automation system is common, regardless of the actual application. This is the case not only in terms of how the initial idea might be developed into a project, but also in terms of how the project should be managed to help ensure eventual success. This chapter is devoted to a description of these processes, and in particular to their application to flexible manufacturing systems. However, the intention is not simply to present a discourse on the various project accounting and planning methodologies which exist. Instead, the content of this chapter has been evolved from these techniques and tempered with experience of the real problems associated with implementing an FMS. The aim is to provide a framework to ensure that all eventualities and project steps are fully taken into account.

Generally speaking, there are five fundamental phases through which major projects of this type tend to pass.

(1) Conceptual Design or Systems Definition
(2) Detailed Design and Functional Design
(3) Actual Implementation (purchase, installation etc.)
(4) Commissioning and Production
(5) Post Start-up Audit

The last three phases of the implementation will, for the most part, be covered in chapter 11, however the first two stages will be covered during the remainder of this chapter.

3.2 Germination of an idea

There is no doubt that if a project which represents a substantial investment for a particular company is to stand a significant chance of being a success, direct management involvement at a senior level is essential. This involvement or commitment will only be aroused if upper management is both interested in, and fully aware of, all the advantages, disadvantages and risks associated with carrying out and not carrying out the project. This will only occur if the initial interest is evolved into a full, though perhaps not necessarily detailed, understanding of the system's capabilities and implications.

The initial interest in the possibility of implementing a flexible manufacturing system could arise from a number of sources. Typical catalysts could be trade journals or technical conferences, or possibly a computer system or machine-tool salesman. Alternatively, it might be the result of an informal social meeting or perhaps even government pressure.

However, if the structure and environment within the organisation do not facilitate the growth of such proposals, then the chances of the idea evolving into a recognised and funded project are somewhat reduced. For example, if it is suggested that an organisation should invest millions in automation equipment while the investment policy is not to commit to any major capital expenditures, there are two battles to be won. First to justify the expenditure, and then to justify the FMS.

Also, the rate of propagation of such an idea is without doubt very dependent on the level at which the idea enters the organisation. If a student apprentice recommends a significant change in a company's manufacturing policy, the suggestion might take somewhat longer to progress than if the Chairman of the Board utters the same words.

Certainly, most flexible manufacturing systems that have been implemented to date have needed a 'system champion'. This is someone who is usually at a reasonably high level within the company. Someone who is able to convince others of the desirability of carrying out a particular project. The 'system champion' may or may not be the idea originator. However, frequently, but not always, this person will become accountable for the implementation of the project, possibly including being responsible for the day-to-day management of the project. But without a 'system champion' both the idea and resulting project are again considerably less likely to be successful.

Regardless of how the initial interest is aroused within an organisation, it must be accompanied by an appropriate depth of understanding of the implications of the project. Usually this will mean less detail for upper management, since it is more concerned with strategy rather than with issues of detail. Nevertheless, all the information is extremely important in its own context. The project is invariably at a very fragile stage, possibly being both politically and technically vulnerable. It is imperative that the necessary information is available in an accurate and understandable form to assist with the formulation of the correct decision regarding the project.

However, it is not only the information itself which is valuable, but also the discipline imposed to make its collection possible. Unfortunately, the data is often quite difficult to acquire (especially if no prior FMS experience exists within the organisation), essentially because of the inherent complexity of the project and the far-reaching inter-departmental implications. Yet this information must be obtained in its entirety before management will be in a position to decide whether or not to pursue the project. Not that a decision cannot be taken without this information! Unfortunately, there are many examples of instances where lack of up-to-date or correct information has not prevented an incorrect decision from being made.

3.3 The content of the Conceptual Design

The way in which an organisation usually obtains this information is by undertaking some form of preliminary analysis. A variety of terms are used as titles for this analysis. 'Feasibility Study' and 'Conceptual Design' are probably among the most popular. The first of these has been used extensively within EEC countries (probably as a direct result of the recent government funding for such named FMS studies). Within the USA, however, the somewhat more optimistic term 'Conceptual Design' appears to be more prevalent.

This type of investigation, which is usually carried out by a relatively small number of people, would have a number of strategic, far-reaching questions in mind:

- Where does the organisation wish to be in, say, 5 years' time with in-house manufacturing?
- Is the investment going to provide the company with strategic advantages over its competitors?
- Will the ability to produce varied products of consistent quality provide leverage in the marketplace?
- Will the publicity involved resulting from such a venture be significantly beneficial?
- Are any industrial relations problems likely to be initiated?
- Is this system to be instigated in isolation or is it the first step within a well-defined long-term manufacturing and automation strategy?
- What are the inter-departmental implications of installing the FMS?

How will it affect the existing management information and production control system?

It is imperative that these strategic questions are both asked and answered at this early stage of the project. Buying an FMS is very different from buying a centre lathe or even a CNC machine. Individual machine-tools, unlike FMS, rarely have a significant effect on the entire manufacturing environment into which they are placed. If this, inevitably substantial amount of investment is not directed optimally the company could well fail to survive. Admittedly, when a company fails it tends to be as a result of a number of adverse influences, the collection of which has caused the eventual demise. The point remains that investing in FMS can be such that if the funds are mis-directed the impact can be sufficiently significant to assist with the collapse.

Of course, the strategic issues mentioned above would also need to be considered in addition to those of a more down-to-earth and detailed nature, such as:

- Why invest at all?
- What are the alternatives to investing . . . now?
- Can the investment be phased?
- Can the project be justified in steps as well as in total?
- How much investment is necessary?
- Why an FMS, why not stand-alone CNC?
- How long would it be before the system is fully productive?
- How much technical risk is involved?
- Is this a reasonable technological step for the organisation as a whole to take?
- What are the operator and training implications?
- What are the maintenance implications?
- If the production from the FMS provides a certain proportion of, say, final assembly requirements, what will be the effect in the assembly shop of having some parts available quickly and others traditionally?
- How much component design analysis should be carried out to ensure that the automation is near optimal efficiency rather than minor piecemeal automation of the existing process?
- Should the parts be redesigned prior to incorporation into the FMS, and if so what are the timing and resource implications of such an exercise?
- How are the components to be selected for the FMS?
- How will the FMS interface with other computer systems (if any) within the organisation, such as CAD, production control etc.?
- How will the FMS integrate with the stores and tool-room activities

already existing within the organisation?
- What are the design options for the FMS?
- Which is the preferred design and why?

Although these questions are far from exhaustive, they do at least go to show both the wide variety and number of questions which must be asked at the very start of the consideration of the project. Even at this stage, it is essential that the project has backing at a level of management which is high enough to ensure that all the necessary information will be made readily available. Typically, much of the information that is required will be located within a number of different departments. Probably not all these departments will be controlled by the same director or manager. Therefore, the co-operation of all the appropriate managers and directors will be needed before the information can be collected easily. Otherwise not only will it be difficult to collect data, but also it might be questionable as to whether the data is either accurate or complete in areas where it should be.

However, while ideally all these questions should be answered in their entirety, at this time this probably would turn out to be a prohibitively expensive exercise. In certain circumstances, therefore, it might well prove possible to settle for a slightly lower level of detail than would be acceptable at a later stage of the project. This would be justified in terms of helping to minimise the cost of developing an outline concept for a particular project, perhaps at the stage when a variety of investment alternatives are being considered. It is difficult to give guidelines as to what is acceptable in these circumstances, and what is not. As intimated above, it is obviously safer to have all the information available that one could possibly desire. However, in real situations this idyllic state of affairs rarely exists, either because of constraints of time or funding. As usual, the 80/20 rule can be applied; it will probably take 20 per cent of the time to collect 80 per cent of the data, and the remaining 20 per cent of the data would take four times longer. Of course, the risk is that a fundamental piece of information is overlooked. This is less likely if experienced people are carrying out the analysis, but could be a serious problem if those carrying out the analysis are not aware of all the questions which should be asked. Ultimately, care should always be taken when drawing conclusions from incomplete data.

An example of such a subject could be the question of consumable tooling. The relevant section in the Conceptual Design would need to include an estimate for the expenditure required to start and maintain production on the new equipment. If the parts are already being manufactured, some knowledge of the tooling requirements will already exist, perhaps the type of tools, their wear rate and a typical unit cost. One might feel that a suitable estimate of the cost of equipping new machine-tools to produce the parts could be obtained by using these figures as a basis for the calculations.

However, if the parts were to be redesigned for the FMS, or produced on different or new machine-tools, this could have a significant effect on the tooling requirements. In this instance, since the analysis is not complete — that is, all relevant decisions about the system have not been made — some certainty level should be attributed to the derived tooling figure. Certainly such a figure is better than nothing, but the reasoning behind its presence should still be made known to the decision-makers.

Certainly, if the Conceptual Design does not have adequate backing from upper management it is unlikely that these questions will be able to be answered adequately. Instead it could well be that inter-departmental arguments are started as to who should influence what part of the design etc. If such a situation arises, the project should probably not even be allowed to proceed, since its eventual result is likely to be an inefficient compromise, and a poor use of company funds.

It is important that when an FMS is installed it is not only the small area of shop-floor on which the equipment is located that is affected, it is in fact more likely to be the whole organisation which will need to be changed. Many people when first embarking on an FMS do not appreciate how far-reaching the effect of installing such a system is likely to be. Experience vindicates this fact, but if the experience is not available it is easy to underestimate the effect that such a system is likely to have. Even when such experience is procured, perhaps on a short-term basis from a consultant, it is not always easy to appreciate the true gravity of the warnings which may be issued.

3.4   Implementing the Conceptual Design

So, having decided that such a study is essential, who should carry out the work? After all, the questions which must be asked are of both a strategic and detailed nature. On the one hand, how does the organisation currently carry out its business and how is this intended to be done in the future? On the other hand, why is one particular tool substrate used in preference to another, and so on.

Also there are a number of significant questions which must be asked concerning the level and type of technology to be used. These questions require highly specialised expertise to provide accurate answers. Unless the organisation is fortunate enough to have people who not only have prior FMS experience but who are also up-to-date on how technology has been developing and therefore know what is and is not technically feasible, outside assistance will be needed. In this case a small elite team should be created to carry out this exercise. This would comprise three or four people from within the organisation and some from outside. These could be a Project Manager (who would probably be reporting direct to say a main-board director), and two or three senior analysts. All these would be experienced and respected members of the company, perhaps representing those departments who would be affected most by the installation of the FMS. Ideal people are those who are technically aware, and who are unlikely to be unreasonably biased towards the company's traditional approach to solving particular manufacturing problems. Innovative thinking at this stage of such a project is highly advantageous.

Finding in-house expertise of the correct calibre may not be easy, but if it is possible this process will not contain so many unknowns and potentially expensive pitfalls as obtaining the desired input of expertise from an outside source might produce. However, if an outside source is to be contacted, it is likely to comprise one or more of the following three possibilities:

(1) Consult an academic establishment.
(2) Try to 'acquire' someone from the marketplace.
(3) Find a competent consultant.

Seeking assistance from academic establishments such as universities and technical colleges frequently has the advantage of being relatively inexpensive and often results in a substantial number of bright enthusiastic students, usually postgraduates, being made available. However, with the best will in the world, academic establishments do not always conform to industrial commitments and deadlines. There are many instances where industry and academia have collaborated extremely effectively, but there are probably substantially more instances of less than happy joint ventures. In many respects this is unfortunate, since most countries have a significant technological resource locked within academic circles and yet it remains difficult for this to be made directly available to industry. However, good production engineering schools have much to offer. The key factor is to know what is possible and what is not, and for both parties not to have unreasonable expectations from such a relationship.

Trying to find someone on the open market also poses problems. People with prior experience of implementing FMS are few and far between and therefore both difficult to locate and expensive to recruit. Although eventually it might be quite justifiable to engage such a person to manage the project and implement the system, it may not be desirable at this relatively early stage of the project. After all, the project has not yet received formal permission to proceed. Obviously, it would be undesirable to increase the company headcount with such an expert, if the analysis concluded that no investment could be justified.

The only alternative remaining is to find a consultant. But how should one be chosen? After all, there are many such consultants offering their services to the market, and yet, as already mentioned, real FMS expertise is rare and difficult to locate. Therefore, it is unlikely that all the consultants are as good as they perhaps would like to be. Inevitably, some will be better than others.

To some degree, it is virtually impossible to know whether one is choosing the best consultant, but certainly there are some ground rules which, if followed during the selection process, should prevent one from making an expensive mistake. After all, it should be borne in mind that the risks of choosing a bad consultant go much further than the cost of the consultant's services for the feasibility study. If the consultant were to be allowed to produce an inadequate FMS design, and if this design were implemented, the system might not even work! This would truly be an expensive mistake. Either way, the consultant's charges are unlikely to be small, and there is little guarantee as to the quality of the work carried out.

The main factors to be considered are:

- To whom within the company will the consultancy be responsible for carrying out the work?
- What real experience does the consultant have of implementing FMS in the appropriate environment?
- How many systems has the consultancy actually been responsible for implementing, and what guarantees as to the performance of the system were given?
- What is the origin and background of the consulting service, is it manufacturing or financial, and what is its depth?
- How many experienced manufacturing people are there within the consultancy service?
- For whom have the consultants carried out work previously (were these customers satisfied and of the same size within a similar industry)?
- What simulation systems do the consultants have available (if this work is not going to be carried out in-house, see chapter 7)?

All these questions are relatively straightforward and it is purely a matter of commonsense to ask them. Probably the single most important issue is that of long-term support of the work carried out. It is to be hoped that the consultants are bringing to the purchasing company the knowledge and experience of how to design, install and operate a sophisticated and reliable system which is more likely to be successful than a design derived from purely in-house expertise.

However, the only sure way to guarantee the external experts' long-term commitment is to find some who are willing to be involved in the implementation of the whole system. Then, any inadequacies within the design can be corrected as a part of an on-going contract.

Unfortunately, this approach can still lead to difficulties since the consultant may just 'monitor' the project, and the customer and subcontractors would have to bear the major part of any increased work-load that results from late modifications etc. To ensure that this situation does not occur, one must try to find a consultancy organisation which is an offshoot of one of the potential suppliers of the system — that is, someone who has a significant long-term interest in the system being a resounding success.

Typically, such a supplier would be a major equipment manufacturer, possibly of say machine-tools, or perhaps controllers. In any event, a manufacturer of a strategically important part of the FMS may be the appropriate source: ideally, someone who is willing to become the single co-ordinating source for the supply of the whole system (that is, a single-source co-ordinator, see chapter 12).

Unfortunately, there are risks associated with even this approach. If, for example, the supplier is a machine-tool builder, one is likely to have the choice of process equipment severely restricted, and indeed the design of the system might not take advantage of technology beyond the scope of that supplier. However, with a true single-source co-ordinator, whose actual product involvement in the system could be relatively small (or perhaps oriented towards the supply of FMS control system), the potential for the eventual user to ensure the system is more optimal for the particular requirements is much higher. Selection of such a single-source co-ordinator permits, for example, the possibility of bringing more influence to bear on the selection of the process equipment, an area of expertise which is likely to be far higher within the user company than within any individual supplier or consultant. Needless to say, such relatively unbiased and experienced system implementers are not plentiful, but they do exist.

Assuming an appropriate outside source of expertise is located, the next step is formally to create a project team within both organisations. Responsibility for particular work areas and day-to-day communications should be established. Ideally the first phase of the exercise should be the customer, ensuring that the consultants are fully aware of the objectives of the system as currently perceived and any relevant constraints on the project. However, this should not be allowed to restrict the approach adopted for the Conceptual Design. After all, the goal of such an analysis should be to outline a manufacturing strategy for the company, with particular reference to the first project. It should not be solely to examine the requirements of this preconceived project in isolation.

This establishment of overall requirements is likely to take between two to three weeks following the signing of a formal contract and the agreed start date. Then there will be some four to six weeks of detailed information gathering (obviously the actual duration will be dependent on the size and complexity of the proposed system). This will be followed by a further two to four weeks during which the data will be assessed and a Conceptual Design developed. This design would then be simulated, perhaps on a PC-based simulation system and probably refined. This process is likely to take a further two weeks. In all, therefore, the Conceptual Design study can be expected to take at least three months of elapsed time to complete. Throughout the whole of the Conceptual Design study, it is essential that the company's project team is fully involved with the work being carried out by the consultants.

Once the consultants have finished their work, a report should be generated. In the main this should be prepared by the consultants and should be presented both in writing and, slightly later, verbally. The verbal presentation is particularly important since it gives upper management an opportunity to question and hence judge the quality of the work carried out. In certain instances it might be appropriate to have a second report prepared independently by the company's project team. This could be as an addition to the consultant's report, or possibly included as a part of the consultants' report. In either case the members of the company's project team should have the opportunity to express themselves freely, without being compromised by the consultants' opinions.

As mentioned above, this process is likely to be relatively expensive, though the costs will probably only represent a small proportion of those of the total project. However, its importance cannot be overstressed. As automation systems continue to become more and more complex, so it becomes vitally important that a sound Conceptual Design is created. This is not only to ensure that the best design is selected, but also to test out the concept and assess how it will fit in with the company's existing plant and long-term manufacturing strategy.

Funding such design studies prior to the implementation of a project nearly always appears to cause companies difficulty, quite unreasonably so. This is presumably because companies are used to simply replacing plant on a like-for-like basis and therefore (perhaps again incorrectly) do not feel that such a study, especially if costly, is really justified. However, the Conceptual Design study should be regarded as an insurance policy. Part of the brief could well be that, given the current situation and the company's long-term goals, how should it invest? Typically a Conceptual Design study will only cost between 1 and 3 per cent of the total cost of an FMS. This is a remarkably small price to pay for peace of mind, so long as this money is spent wisely.

3.5 The Project Team

On the basis of the results of the Conceptual Design study, the management of the company is now going to need to make some important strategic decisions:

(1) Is the company going to proceed with implementing the FMS?
(2) How is upper management going to maintain its involvement?
(3) Who is going to manage the project?

If the company is going to proceed, then certainly the appropriate financial provisions will have to be made both in total and at the necessary rate, as should have been identified within the Conceptual Design.

Since the investment is almost bound to be substantial and of a strategic nature it is likely that upper management will want to be involved, though probably not on a day-to-day basis, but at least on a monthly basis, particularly during the earlier formative stages of the project. One of the best ways of ensuring that this occurs is by forming a Steering Committee. One member of this would be the Project Manager, the remainder might be company directors. An alternative approach would be simply to have the Project Manager reporting to a single director. This is acceptable if the director is sufficiently influential, but the multi-department coverage automatically provided by the creation of the Steering Committee is preferable. The main purpose of the Steering Committee is relatively straightforward. It is to support, guide and review the work being carried out on the project and ultimately to take collective responsibility for the eventual success or failure of the FMS.

So, with a positive decision resulting from the Conceptual Design study, funds available and a steering Committee formed, the next task is to find a Project Manager. As mentioned previously, if somebody exists within the company who is appropriate for this task, all well and good. If the company has to compete on the open market for a Project Manager the task could be both long and expensive.

Ideally the Project Manager should be completely familiar with recent advances or in relevant spheres of technology, and in particular should be familiar with sophisticated computer systems, microprocessor control techniques, machine-tools, robotics and work handling. Prior successful project management experience is essential, preferably related to FMS implementation.

As mentioned before, such people tend to be difficult to find, which is not surprising given the above list of desirable accomplishments. This unfortunate situation is mainly a result of the newness of the technology, but is also in part an outcome of the 'single subject' structure of most manufacturing related courses at universities etc. (see chapter 12). Certainly where FMS is concerned a broad spectrum of experience is essential. However it is now apparent that computer control systems tend to represent the heaviest and probably the most fundamentally important workload. Therefore, it is probably desirable to have a Project Manager whose expertise is biased towards this discipline, although to some degree this decision will have to be made in the light of the requirements of the particular system being considered.

Among other matters, the Project Manager will be responsible for the following:

- The day-to-day specification and progress of the project, starting with the systems design effort and eventually finishing with the Post Start-up Audit of the system.
- Review and approval of all major system design decisions.

- Development of a detailed systems implementation schedule and ensuring that the project stays on schedule.
- Preparation and presentation of periodic reports to the Steering Committee.
- Staffing for all phases of the project.
- Making recommendations to the Steering Committee when strategic decisions are required.
- Infusing and maintaining good working relationships between all the personnel involved. (*Note*: this would include the subcontractors concerned with the progress of the project.) This is a particularly important and often underestimated task, made more important by the complex mix of technologies within an FMS which frequently need to be supplied by a variety of manufacturers.
- Ensuring that the FMS will integrate into the existing manufacturing environment.
- Training and education of the operators and maintenance staff to be associated with the system, together with anyone else it might affect, such as raw material suppliers, operators, upper management etc.

Once a Project Manager is in place, it will be possible to progress to the next phase of the implementation.

3.6   Detailed Design of the FMS

The Detailed Design phase of implementing an FMS is, in many respects, the most crucial phase of the project. It is essential that the design is specified with considerable accuracy and devoid of areas of uncertainty before the bulk of the equipment and software are purchased. While it might have been possible to approximate some data during the Conceptual Design, such incomplete data is now incompatible with successful project execution.

However, it is likely that some long lead time items will have to be ordered prior to full completion of the design, for example, some of the machine-tools. With items such as these, it is unlikely that strategic issues such as the choice of the supplier and the fundamental design of the equipment itself are likely to change as a result of additional work being put into the design. Possibly some of the ancillary equipment such as pallet changers and tool carousel capacities might well change. But this should not pose a particularly significant problem to the supplier, especially if the suppliers have been made fully aware of the situation.

The work that needs to be done during the detailed design phase is essentially a more in-depth analysis of the work started during the Conceptual Design study. Some of the factors which typically would be considered are as follows:

- Which components should be manufactured and why?
- Can they be redesigned to simplify production?
- What is the life of the products now?
- Are the production requirements likely to change?
- What are the volumes of the components required?
- What batches sizes should the system produce, and is this appropriate for the next operation, say assembly?
- How important will it be to minimise set-up times and what will be the cost effect?
- What are the process cycle times?
- How will swarf and coolant be removed?
- What are the tooling implications? How many at each workstation? How and by whom will they be refurbished?
- What are the quality control implications — is inspection to be post cycle or post process, manual or automatic, on the workstation or off?
- How much quality data will need to be archived?
- How long are the part programs likely to be and how will this affect the transmission rate on the communications network?
- How will work schedules be entered, adjusted and optimised?
- How will the system recover from failures? What level of failure resilience and graceful degradation should be built into the system?
- How will component definitions and process routes be entered into the system?
- How will the system interface to existing manufacturing systems — CAD, production control, inventory, marketing, financial etc?
- What are the material handling implications — are robots needed, conveyors and/or AGVS?
- Is there a requirement for vision or voice recognition systems?
- How many operators will be needed? What will be their skill/training requirements?
- How many shifts per day will the system be operated and on how many days per week?
- How will the system impact on the current maintenance capability of the company?
- Where will the system be located — are special areas required, or foundations for heavy machine tools, air conditioning for computers?
- Is a new building required?
- Is the system to incorporate any existing equipment, and if so will it need to be modified?
- Are any industrial relations problems likely to be created by the implementation of the system?
- How far reaching will be the impact of the system given current departmental/managerial boundaries?

With all these and many other questions in mind, the Project Manager will have a number of tasks which require urgent attention. The first of these will be to select a Project Team which is appropriate to the nature of the project, and is particularly strong in any areas in which the Project Manager is deficient. As is obvious, during this time the Project Manager will also have to become extremely familiar with the results of the Conceptual Design study, possibly learning about the current manufacturing policy and techniques. Certainly the Project Manager will need to become aware of the political environment within the company, largely because of the multi-disciplinary nature of FMS projects. One way of easing any potential problems in this area is to ensure that the Project Team includes a representative of the relevant departments. This may not always be possible, but even if it is, the Project Manager still has to ensure that these people are allocated to the project on a full-time basis, and are directly responsible to the Project Manager.

Since a Project Team should have been formed to produce the Conceptual Design, it may well be desirable and prudent to select some if not all the people involved at that time. It is likely, however, that some changes will have to be made. Some additional personnel will probably be required, and some of the people previously involved may not be available on a full-time basis for the duration of this part of the project.

The Project Team should not consist of more than six people, including the Project Manager, since five is generally accepted as the largest number of people which can be effectively managed by one person. A typical mix of experience would be as follows:

Project Manager   — Supervisory
                   — Personnel and budget
                   — Negotiating/relations
                   — Computer systems
                   — Process and materials
                   — Work handling
Project Engineer 1 — Computer systems, hardware and software
Project Engineer 2 — Process technology
                   — Components
                   — Products
Project Engineer 3 — Work handling
                   — Components
                   — Products
Project Engineer 4 — General engineering
                   — Site preparation
                   — Draughting
                   — Installation

The structure of this team and how it would interface to the Steering Committee would typically be as shown in figure 3.1.

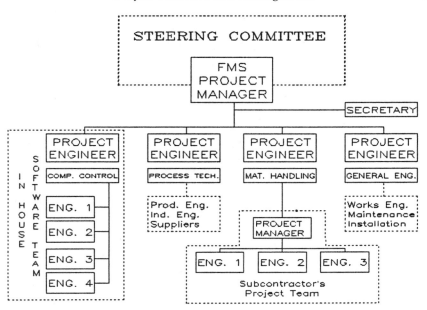

Figure 3.1   Typical FMS project management hierarchy

Obviously, depending on the type of project involved and the amount of work being carried out 'in-house', the mix of experience would need to be adjusted as appropriate. However, it is always desirable that the Project Team is kept small and is not physically dispersed. Both these factors will facilitate frequent and detailed exchanges, the cross-fertilisation of ideas, the sharing of problems and a thorough understanding of project progress and status.

If, for example, the control software is to be written by an in-house data processing department, there might well be a software development team of, say, a further four or five people assigned to the project. In this case, it would be sensible to have the appropriate Project Engineer as the leader of this software team. All the other members of the software development team would also need to be made to feel members of the FMS project as well, but technically they would normally report to the Project Manager via the Project Engineer — that is, their manager. This will ensure that at every level of the project an individual manager does not ideally have more than five people reporting directly. This greatly improves the efficiency of the managerial task and also helps to ensure that all aspects of the project are closely co-ordinated.

Close communication between all the people assigned to the project, regardless of the organisation from which they originate, is extremely important. This is one of the fundamental goals which is being targeted by the above management structure. For example, the control software will have an impact on virtually all aspects of the FMS. Therefore, a considerable amount of communication is required between the appropriate software team member and the project's equipment experts if the relevant parts of the system are to be co-ordinated. If the main work transfer mechanism was to comprise a sophisticated computer-controlled conveyor system, it is essential that all those involved in the development of the conveyor's control software, led by their team leader, are in virtually continuous contact with those developing the conveyor hardware.

So far, this discussion has put forward suggestions on how an FMS project team might typically be put together. However, to date it has tacitly been assumed that the main Project Team is within the eventual user's organisation. This might not be the case. For example, the user might be working closely with a single source co-ordinator. If this was the case, the user could reasonably expect to see a Project Team hierarchy along the lines outlined above established within the system co-ordinator's organisation, and a much reduced one within their own organisation. It would be highly desirable for a similar project team to be established within the user's organisation, but obviously the workload on the project team members would not be substantial. Conceivably, the number of members of the team could be reduced by the sharing of responsibilities, the bulk of which would be the 'supervision' of the co-ordinator's team members activities within the user's organisation and

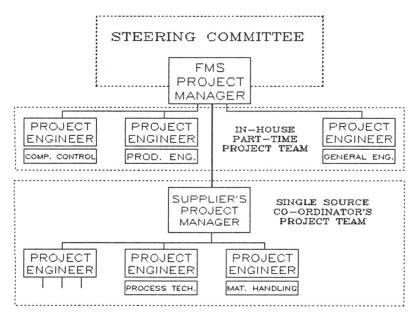

Figure 3.2   An alternative FMS project management hierarchy

the joint specification of the system's production and operational requirements. It is likely that these people would not need to be devoted to the project on a full-time basis, but ultimately only for as much time as is needed to ensure that the system being supplied is what the customer has requested. In this case the Project Team architecture could well be as shown in figure 3.2.

Regardless of the final form of the project management hierarchy, it is worth bearing in mind that ultimately it is not the technological equipment within an FMS which makes the FMS a success; it is more the way in which the various technologies interface and integrate with each other, and this is dependent on how well the various people involved with the FMS communicate with each other.

If there are any significant personality clashes within the Project Team, it is desirable to find some means of eliminating these as soon as possible. It is imperative that this group of people works together at all times as a small, dynamic, responsive, tightly co-ordinated and, above all, committed team. If, for whatever reason, this proves not to be the case, then the chances of eventual success of the system are much reduced.

3.7 Design Reviews

It is also highly desirable to have frequent reviews throughout the duration of the project: both informal project meetings (which should be on a regular basis, although the frequency of the meetings would vary according to the stage of development of the FMS) together with the more-detailed and formal, less-frequent Design Reviews, which would also be highly beneficial to external participants. It is unlikely that, for much of the project, the informal meetings would need to occur more frequently than every two weeks. But not to have a meeting for more than two months would be most undesirable. In either case, such meetings are of value for a variety of reasons:

(1) All members of the team can report their individual progress.
(2) Team members are able to raise interface issues with the other members, as required.
(3) The Project Manager is able to assess overall progress, allocate individual workloads, and inform the group of any relevant Steering Committee decisions.
(4) A free flow of ideas is encouraged from one discipline to another.
(5) A forum is created where any project related problems can be raised.
(6) The Project Team is forced to pause from the hectic addressing of the detailed requirements of the system, and to rigorously consider the integration and overall direction of individual efforts.
(7) A good team spirit is generated.

In fact, detailed Design Reviews are of considerable value to any complex project, almost regardless of the stage of its development. While the goals of such a Review, in terms of the particular project being considered, are relatively introspective as expressed above, the goals from the point of view of the entire organisation involved, and particularly to those participants external to the project, whose fresh perspective can be invaluable, are usually more far reaching. Yet both can be accomplished simultaneously, for example, by including the following considerations:

(1) Maximising return on all related projects.
(2) Inter-departmental sharing of design and development effort.
(3) Ensuring compatibility of resultant systems.
(4) Minimising the risk of failures and false starts.
(5) Minimising expectation uncertainties

The topics which would be considered within such reviews would alter depending on the stage of development of the project. Certainly there would be common elements, but the emphasis on these is likely to change slightly. For example, during a Conceptual Design Review the topics covered would include:

- Expectations — requirements, capacity, cost, timing etc.
- Conceptual planning
- Trade-off considerations
- Preliminary design
- Type of parts to be manufactured
- Modes of steady-state operation
- Number and types of equipment
- Group technology considerations
- Layout
- Computer systems
- High technology and/or high risks areas
- Process and associated technologies
- Quality control considerations
- Operator skill requirements and training
- Justification including simulation
- Documentation (design, operation and long-term support)

During a Detailed Design Review the topics covered would include:

- Development of conceptual considerations
- Analysis of Detailed Design specifications
- Manufacturing systems design
- Control systems design
- Cost and timing implications

- Integration with existing systems
- Loading factors and simulation results
- Identification of changes and problems
- Review documentation

Although overall economies of resources are most easily identified during the Conceptual Design, during the Detailed Design the ultimate success or failure of the system is decided. Similarly, whether the system is on time or not depends on how carefully the tasks, and their respective workloads, are forecast and scheduled at this time.

There are three rules which, if rigidly adhered to, will assist considerably in attaining the project goals:

(1) The Detailed Design must cover every aspect of the FMS; everyone associated with the project should be fully aware of the scope and extent.
(2) No major commitments should be made on the components of the system before the Detailed Design is complete (particularly regarding software).
(3) Any changes that have to be made to the Detailed Design, after it has been issued, should be rigorously documented. Indeed, a design change procedure should be included within the Detailed Design (*Note*: this is particularly important, since most suppliers have quoted against the contents of the Detailed Design.)

Finally, experience shows that the Detailed Design phase of the project should not be hurried. Time spent in ensuring that the design is comprehensive is time very well spent. If this phase of the project takes longer than expected, then so be it. Better to be slightly late with a working system than on time with one which does not work.

3.8 Project timescales

Obviously it is virtually impossible to be specific about the timescales, individual workloads and budgeting implications of implementing a particular FMS since there is so much variation between different installations in terms of both size and complexity. However, it is possible to make some general comments which should be of value to potential FMS implementers.

The major elements of an FMS project are the preparation of a Conceptual Design and obtaining management commitment; Detailed Design; implementation and procurement; installation, commissioning and test. Items of major equipment might well have relatively long lead times associated with their supply, as indeed will the control software. Even after the machine-tools have been installed, there is a considerable amount of commissioning, testing and debugging that needs to be carried out before the system may be considered operational. So, unless the FMS is either small or being produced from largely existing equipment etc., it is unlikely that the project duration will be much less than three years. These 36 months would typically be allocated to the various phases of the implementation as shown in figure 3.3.

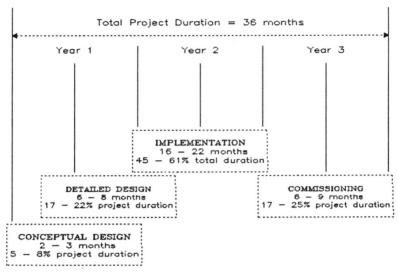

Figure 3.3   Typical FMS project timescales

In many respects the preparation of the Conceptual Design is a unique task. It is the only phase which is entirely separate from all the others. This is necessary since only after the Conceptual Design has been developed within the long-term manufacturing strategy of the company as a whole, is it possible to obtain the managerial commitment to start formally the project and progress to the Detailed Design phase. However, once this has started, the remaining phases become slightly less distinct and tend to merge into one another.

For example, the Detailed Design task shown represents 17–22 per cent of the total project duration. It does not include sufficient time to develop a detailed design of all the control software necessary (assuming that this is going to be written especially for this FMS). However, it does include enough time to specify the overall software architecture, and detail the relevant equipment interfaces. Similarly, with 'commissioning', much of the equipment and software will have been tested in isolation prior to delivery to the FMS. Indeed, such testing should be encouraged and be as comprehensive as possible. However, there still remains the collective testing and debugging. It is for this reason that the border between these latter phases is shown to be less distinct.

The following list gives some indication of typical project milestones against which a project schedule could be prepared and monitored: detailed tasks into which the project should be subdivided at an early stage.

- Management Commitment
- Select Project Manager
- Establish Project Team
- Develop project management tools
- Select components
- Develop computer simulation
- Establish financial justification
- Select processes and equipment
- Identify suppliers
- Select material handling equipment
- Identify suppliers
- Develop outline systems design
- Develop software functional design
- Establish computer hardware specification
- Identify suppliers
- Carry out detailed software design
- Code and test modules
- Test integrated software modules
- Prepare site
- Install process equipment
- Test process equipment
- Install material handling equipment
- Test material handling equipment
- Integrate and test
- Prepare part programs
- Install main computers
- Install and test software
- Test communications system
- Test software and process equipment
- Test software and material handling
- Integrate and test
- Recruit and train operators
- Train maintenance personnel
- Start production
- Integrate FMS within factory
- Full production
- Post Start-up Audit

While this list is far from being complete for any particular FMS, it should nevertheless give an indication of the level of detail which needs to be considered. Time spent at the beginning of the project ensuring that such a list is comprehensive and accurate will be well spent. However, this information would probably be presented in summary form to upper management, perhaps as shown in figure 3.4.

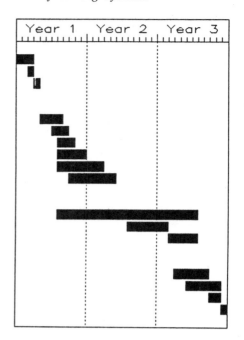

CONCEPTUAL DESIGN:—
Feasibility Study
Management Commitment
Establish Project Team

DETAILED DESIGN:—
Select parts
Process selection
Material handling selection
Order equipment
Identify systems requirements
Order computer systems

IMPLEMENTATION:—
Develop system elements
Site preparation
Install equipment

COMMISSIONING:—
Subsystem testing
Integrated system testing
Gradual production start—up
Full production

Figure 3.4　A summary of a typical FMS implementation plan

Even though the durations of the various tasks will be application-dependent, diagrams of this nature are invaluable for both planning the project and monitoring its progress. However, since the duration of the tasks is likely to change quite frequently, it is highly desirable that the process of creating and updating these project plans is fully computerised.

If, for example, a microcomputer is being used for the bulk of the simulation of the system, there is no reason why this hardware should not also be employed to keep track of the project with one of the many project planning packages that are currently available. For example, Superproject, Total Project Manager and Pertmaster, to mention just a few. Their ability to facilitate the input, modification and update of a complex PERT network or critical path analysis is exceedingly valuable to the busy Project Manager. In addition to carrying out these basic functions, many of these project planning packages have the ability to calculate both the resource requirements and the financial implications of input project plan. This facility is particularly helpful when a change occurs in the project. Very quickly the Project Manager will be able to assess the timescale, resource and financial implications of the change.

3.9 Project budgeting and workloads

Whereas the overall capital requirements of implementing the FMS will be fixed relatively early in the life of the project, the cashflow implications will be less certain. They will be affected most significantly by the billing requirements of the major equipment vendors — that is the suppliers of the machine-tools, computers, material handling equipment etc. Software development is likely to be a fairly constant cost burden on the project, although it should be borne in mind that costs for these tasks might rise slightly if specialised assistance is required on a temporary basis during perhaps the Detailed Design phase of the project. Figures 3.5 and 3.6 show, respectively, examples of the output which might be expected from a typical microcomputer project planning package when printing out monthly actual and cumulative project expenditures. Meanwhile, figure 3.7 shows the cumulative cashflow which might be expected to occur within an FMS project which proceeds as shown in figures 3.3 and 3.4. The major assumptions used to prepare this particular chart are shown below:

	Percentage of project budget
Conceptual Design	3
Detailed Design	17
Site preparation	5
Process equipment	22
Material handling	17
Computer hardware	10
Computer software	20
Training	3
Commission, test and debug	3

It is likely that the workload of the Project Team will be greatest during the earlier section of the Detailed Design and the final stages of implementation. Due allowance for this fact should be made when considering holidays etc. At these times especially, it is essential that all the relevant personnel are readily available. A breakdown of the project workloads which would typically be expected is essentially as shown in figure 3.8.

The purpose of this chapter has been to introduce the reader to the project management complexities involved in implementing a flexible manufacturing

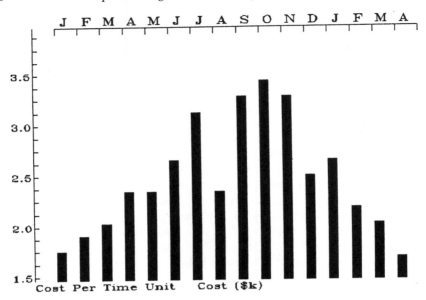

*Figure 3.5   Monthly project expenditures as generated by a microcomputer
project planning package*

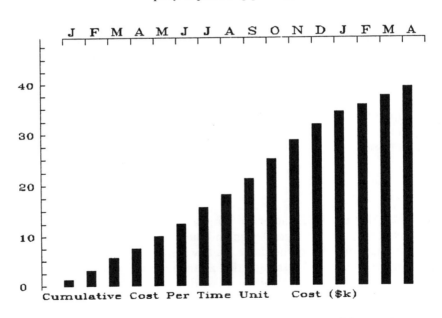

*Figure 3.6   Cumulative project expenditures as generated by a microcom-
puter project planning package*

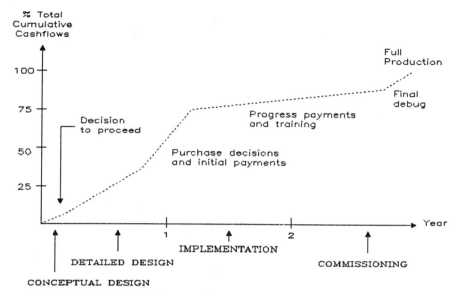

Figure 3.7   Typical cumulative FMS cashflow implications

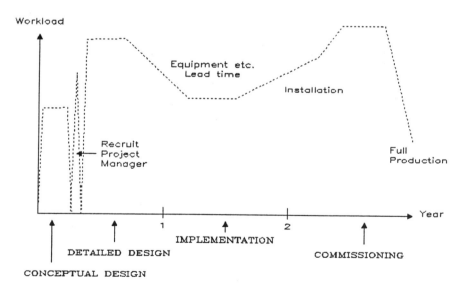

Figure 3.8   Typical FMS project workloads

system. In fact, much of what has been said applies equally well to a wide variety of automation and other projects. Chapter 11 is devoted to a discussion of the later phases of implementing an FMS, which have not yet been covered.

4 The Components

4.1 Introduction

Surprisingly, it appears that many people do not appreciate how funda-mentally important the choice of both the components produced and the processes used, are to the eventual design of the system. It is even more surprising how few people appreciate how the components, products or processes should be selected.

This chapter is intended to serve two purposes. First, to give an indication of how the above analysis should be carried out, and second, to act as an introduction to the following two chapters on production processes and material handling.

4.2 The importance of the components selected

The reason why the selection of the set of components to be manufactured is so important is really quite obvious. Virtually all the design of the system, such as machine tools, material handling systems and even the computer system, is completely centred on the production of the selected parts. The flexibility which needs to be designed into the system will have a considerable impact on both the complexity and cost of the FMS, and many design features are dictated almost entirely by the components chosen to be manufactured. Of course, at this stage of the analysis whether it is individual components which are to be manufactured or completed products is not of particular significance. During the later stages of the design this will not be the case.

Unfortunately, in the past, few companies appear to have appreciated the importance of the parts selected to be manufactured by an FMS. Instead, perhaps a situation of under capacity has developed within a certain section of a factory, and, in place of reacting in a more traditional manner by purchasing additional machine tools (which admittedly might be a route equally fraught with pitfalls, albeit of a different type), it is decided that the opportunity should be taken to develop an FMS. It could well be that this reaction is prompted by the availability of government grants or perhaps that one's competitors are seen to be investing in similar technology. Either way, little credence is usually given to the strategic nature of the investment, though to some extent it appears that this situation might be changing.

4.3   A place to start

After having made the decision to invest what will almost inevitably turn out to be a significant amount of capital in an FMS, a number of questions should be asked. The order in which these are asked is likely to vary somewhat, depending on the application environment, but certainly, one should not select some components at random, as such a course of action is likely to cause many problems later. Instead it is far more sensible, if not essential, for appropriate representatives of the organisation first to stand back from the whole of the manufacturing scenario and ask the question: 'if a substantial amount of capital is available to be invested in manufacturing, where is the most beneficial place to allocate that money — machining, fabrication, assembly etc. — and in what form?' Such a question should be asked relative to both the short term and the long term.

As far as the long-term implications of this type of project are concerned, it is likely that people who are familiar with the manufacturing technology of the company, and who also have a knowledge of the company's future product strategy, will be best suited to this task of identifying, with a reasonable degree of confidence, where the funds should be allocated. With these suggestions as guidelines, a detailed analysis should be instigated for confirmation. This analysis should be far reaching and cover all manufacturing areas, while concentrating on those identified previously

As was suggested in chapter 3, it may well be desirable to have an outside organisation carry out this essentially strategic product and manufacturing analysis, to provide the opportunity of generating a fresh perspective. The old saying 'not able to see the wood for the trees' often applies to all types of businesses, and particularly to FMS during the Conceptual Design phase.

It could, of course, be that the whole project to develop an FMS has arisen because the company has decided to launch a new range of products which require entirely new component designs to be manufactured. If this is the case, then the situation is simplified, in so far as component designs, processing routes and equipment etc. are all likely to be variables which can be fixed (at least to some degree) as the design of the FMS progresses. On the other hand, there are likely to be so many variables that it will be difficult to know where to start. Then one of the dangers is that so many decisions have to be made so quickly that mistakes are made, and people are upset. This situation is likely to be further complicated if the FMS is to manufacture some parts which are already being produced.

If many of the components are new, then the design engineers must work jointly with the manufacturing engineers to develop an optimum compromise between design requirements and ease of manufacture. Indeed, such a process should occur even if the components are already in production. Once substantial agreement has been reached, the designs and processing requirements of the parts should be fixed and the evolution of the FMS design continued from the appropriate point as if all the parts were already in production.

4.4 Selecting the components

Typically many, if not all, of the parts to be manufactured will already be in production. The method by which parts should be selected is to apply two types of analysis. First, a Pareto or relative importance analysis is needed, essentially to establish which parts are the most important. Often 20 per cent of parts may represent 80 per cent of the business requirement. Of course, the term 'most important' will have to be interpreted in a manner appropriate to the particular application. For example, it could mean:

- The most valuable parts
- The most difficult parts
- The most commonly demanded parts
- The parts for which the design is most stable
- The parts for which the design is least stable
- The most production-intensive parts
- The most labour-intensive parts
- The most accurate parts
- The parts which are most carried in stock

In fact, probably the safest route is to assume that all these factors are of importance, perhaps with a weighting factor relevant to the particular environment. After having ranked the parts, perhaps within a number of analyses, it will then be possible to see which appear most frequently within most analyses. This type of analysis is relatively straightforward to perform, once the data is available. However, obtaining the data may well prove to be a lengthy task. Certainly, the whole analysis procedure is likely to be speeded up considerably if a computer, possibly a desktop microcomputer is used to carry out the mundane sorting and iterative numerical analysis.

Having identified a set of sufficiently important parts, the next stage of the analysis should be to apply group technology principles. Parts should be grouped together according to common features such as:

- Processing requirements
- Process routes
- Material handling requirements
- Physical characteristics

- Similar materials
- Part families comprising assemblies

Such an analysis may be considerably simplified if a sophisticated part numbering system is already in use within the company. Such systems are frequently selected in such a way that relevant characteristics of the components, such as size, material, accuracy etc., are denoted by various digits comprising the part number. Therefore anyone examining a particular part number is immediately able to deduce a considerable amount of information about the part concerned. Such systems can also be applied to, for example, the tools which are used within the factory. Of course, depending on the number of features to be classified, the string of numbers can become quite large. Some of the better-known computerised systems accommodate up to as many as thirty-two digits, while some companies in the USA have part number strings of forty-two characters. In general, however, ten to sixteen characters is not an unmanageable string. If it transpires that no such system is currently in use within the company, it is probably time one is installed, in which case this would be an excellent way of beginning the FMS investment. An example of such a part coding system based on the MICLASS system is shown in figure 4.1; this is used by General Electric (USA) in some of its plants.

Digit	Significance	System
1 2 3 4	Basic geometric form	Standard MICLASS
5 6 7 8	Primary dimensions	
9 10	Tolerance and finish	
11 12	Material	
13 14	Batch size	
15 16	Secondary dimensions	
17 18	General manufacturing data	
19 — 27	Additional information	MICLASS3

Figure 4.1   An example of a typical part coding system

Similarly, if the company concerned is operating a sophisticated computer aided process planning system, much of this type of information may well be readily available, purely by instructing the computer to sort data in an appropriate manner.

From these analyses a number of sets of components will be generated, each of strategic importance to the company. It could well be that the manufacture of some of these groups is already well in hand, in which case one would proceed to the next group. Eventually, however, the ideal candidates will be filtered out, probably in terms of a long-term need and an inability to address that need adequately. Examples of some typical groups of parts which might result from such an analysis, both cylindrical and cubic, were shown in figure 2.3, parts (a) and (b) respectively. From these sets or groups, for various reasons, it might be appropriate to eliminate certain parts. For example, while most members of the group might share a particular processing route, one part might not, and it would not be unreasonable to remove this part from the group. Similarly, a part from another group which perhaps only needs most of the main group's processing capability may be included, since perhaps it completes a set for assembly. Analyses such as these should be carried out during both the Conceptual and Detailed Designs, perhaps rather more thoroughly during the latter of these.

4.5 Consequential system implications

So, with the parts for the FMS selected, though perhaps subject to some fine tuning, many of the features of the flexible manufacturing system are automatically fixed, either entirely or at least by having significant design guidelines established. For example, given that the parts are selected, the FMS designers will now be able to quantify the answers to a number of particularly significant questions:

- The current and future volume requirements?
- The current and proposed batch sizes?
- The current and proposed processing routes?
- The current machine-tools etc. used?
- The set-up times on the machines?
- The jigs, fixtures and tools used?
- The frequency of replacement of tools?
- The cycle times of the part on the machines?
- The physical type of the components — that is, whether rotational, cubic, light, heavy etc?
- The current work handling techniques employed?
- Work handling cycle times?
- The current number of direct and indirect operators required?

- The level of skill of the operators used?
- Current scrap rates?
- Typical batch throughput times?
- The amount of work in progress inventory currently required?
- The inventory levels of raw material and finished parts?
- Raw materials used?
- The value of the raw material and finished parts?
- Accuracy requirements, inspection level required?
- Current production problems?
- Life of the product, likely successors?
- Ability to change the design?
- Ability to change the manufacturing process?

All these questions should be asked at least of every component being considered for production within the FMS, and ideally of a few of those which were nearly selected, in case it does become necessary to exchange a subset of the parts for some reason at a later date. In certain circumstances it could well be that additional questions resulting from, for example, the particular manufacturing processes employed, should also be asked. All the answers should be carefully catalogued and entered in an easy to read form, in a 'Component Data File' and also in a 'Processing Requirements File', though it will not be possible to complete this fully until the manufacturing processes have been selected, a topic which will be discussed in detail in chapter 5. Figures 4.2 and 4.3 show examples of such data sheets, and how this summary information would typically be cross-referenced to the more detailed data.

As mentioned previously, it is sensible to computerise this data as soon as possible, perhaps using one of the many reputable database systems available. In fact, it is highly desirable to standardise on a particular software package or set of packages for use by all members of the project team. If one of the increasingly popular integrated packages is used, the whole Project Team can interchange data and be provided with the ability to combine database, word processing, spreadsheet and telecommunication facilities. This will facilitate the substantial amount of analysis which will be needed, and the communication of the results, both at this stage and later. While some time will be needed for training and the setting-up of standard data formats, the benefits which will be reaped later with the ease in which data can be prepared, analysed and presented, will be substantial.

4.6   Some strategic issues

Before the implications of the detailed part characteristics can be assessed, some of the more strategic questions should be asked, since they could well be influencing the answers to the ostensibly more straightforward questions. For example, the final three questions, concerning product life and design

ALL—PARTS—MADE—FAST LTD,
Plant 3 — Flexible Manufacturing System Project

| PART DATA | Title: PD1 | Ref to: PD/R1 | Date: |
| | | | Author: |

Part No: A123456—001 Description: Reverse idler shaft

Data type	1986	1987	1989
Part type	Shaft	Shaft	Shaft
Geometry	Rotational	Rotational	Rotational
Material type	EN8D steel	EN8D steel	Plastic
Dimensions (HxWxD)	1 x 1 x 10	1 x 1 x 10	1 x 1 x 10
Weight	1.5	1.5	0.5
Accuracy	Medium	Medium	Medium
Raw material cost	1.5	1.1	0.6
Finished part value	5.5	5.5	5.2
Average value	3.5	3.3	2.9
Annual volume	300	400	600
Batch size	50	50	75
Throughput time (wks)	3	2.5	1.0

etc. ....

ALL—PARTS—MADE—FAST LTD,
Plant 3 — Flexible Manufacturing System Project

| RELATED DATA | Ref: PD/R1 | Date: |
| | | Author: |

Name: A123456—001 until start of 1989   Sheet ...... of ......

Comments: In 1989 shaft will be made from plastic instead of steel, see PD2 and PPRD2 for new data.

Units:
Dimensions in mm
Weight in kg
Values in decimal pounds

Assumptions:
Annual volumes extracted from 1986 5 year plan (ref. .....)
Batch sizes provided by ....

Conclusions:
 etc. ....

Figure 4.2   An example of a part data sheet

ALL—PARTS—MADE—FAST LTD,
Plant 3 — Flexible Manufacturing System Project

PART PROCESSING	Title: PP1	Ref to: PP/R1	Date: Author:

Part No: A123456—001 Description: Reverse idler shaft

Batch size: 50	Material: Steel EN8D	Sheet .... of ....

Op. No.	Description	Location	Duration (min.)			Comments
			Set-up	Cycle	Batch	
10	Check material	Stores	O	1	50	Visual check for imperfections
15	Transport		O	30	30	Via fork—lift
20	Rough turning	P203 M/C	25	1.5	100	Skilled setter, 124 jaws, each operator 3 m/c
	etc.					

ALL—PARTS—MADE—FAST LTD,
Plant 3 — Flexible Manufacturing System Project

RELATED DATA	Ref: PP/R1	Date: Author:

Name: A123456—001 until start of 1989   Sheet ...... of ......

Comments: In 1989 shaft will be made from plastic instead of steel, see PD2 and PP/RD2 for new data.

Assumptions:
Stores location at start of job assumed as goods received
Inter—operation transport assumed throughout to require forklift
Process locations as shown in layout X103 (1985)
Process comments supplied by ....
Cycle times supplied by ....
New cycle times (marked *) are calculated on the basis of ....
Time durations expressed in decimal minutes

etc. ....

Figure 4.3   An example of a part processing data sheet

stability over time, are of considerable importance, since it is likely to take years rather than months before the FMS is in production. In general, therefore, it is impractical to consider components which have a life of less than, say, twice the implementation period of the FMS (unless the tooling and other set-up costs are not significant). Admittedly, many consumer markets are highly volatile, and hence some products do have extremely short lives, maybe of the order of only three or four months. If the FMS is being considered for some of these types of product, then maybe the current components can only be used as a guide to what the eventual manufacturing requirements of the FMS should encompass. In such an instance the FMS designers will have to take great care to ensure that adequate flexibility is built into the system to accommodate likely changes and, of course, these are unlikely to be easy to define since they are obviously of a fairly unpredictable nature. In this case a considerable amount of risk is therefore involved. Flexibility costs money. However, if the correct amount of flexibility is not built into a manufacturing system, its life (and that of the operating company) may be prematurely ended.

Having shortlisted a component or component family for inclusion within the FMS manufacturing capability, the next issues to be addressed are whether any advantage could be gained by altering the current design and/or the manufacturing process, perhaps by simplifying the design and hence making it easier to assemble automatically, or perhaps by reducing the number of operations required to manufacture the part. Conceivably, this might not be acceptable if, for example, the part has not been designed by the manufacturing company (perhaps the FMS is being installed by a true Job Shop). Alternatively, the part might be a component within a design-sensitive product, such as ammunition, where frequently Ministry of Defence approval must be obtained before a design or process change may be made. However, even in these circumstances, the advantages to be gained in terms of system integrity could well outweigh the disadvantages associated with obtaining authority to make the changes.

One temptation which must be avoided is that of simply automating the current manufacturing process. Obviously the current or 'traditional' way in which a part is produced will be dependent to a significant degree on the resources (people and equipment) available to the company when the part was originally designed. Since many designs tend to be long term, while manufacturing technology advances and provides improved production methods, unless the design is for some reason (such as installing an FMS) reviewed, then these potential benefits are not secured. Possibly the part manufacturing processes will occasionally be changed, for example to accommodate the arrival of some new equipment. But unless the process is reviewed in its entirety, it is likely that old and possibly inefficient methods of manufacturing will be maintained. In fact, the manufacturing process currently used within the company may well be ludicrously inefficient, having been justified purely on the basis of tradition and equipment availability.

Obviously, while it is easy to say 'that's the way its always been done, so why change now!', it is far from the best way to operate anything. When implementing an FMS the ideal opportunity exists to observe the production environment from a new perspective, and install the best possible manufacturing processes for the parts. Who knows, it might even be the beginning of a new tradition! Hopefully, however, it will not take the arrival of an FMS project to secure an appraisal of the company's manufacturing procedures. If it does, it may be that the FMS is only the start of what needs to be done to convert the company into an efficient manufacturing organisation.

4.7 Overall processing implications

The next step in the FMS component analysis is to consider the impact of the manufacturing processes on the design of the system. This should be considered initially with regard to the existing design and processing requirements, and later by establishing how the part would be manufactured in an ideal environment (a topic which will be covered in greater detail during the next two chapters). After this, the impact of the results of this analysis on the overall design of the FMS should be established. No doubt compromises will have to be made, especially if some of the existing plant is to be used within the FMS. However, an appreciation of the magnitude of the gains which could be achieved by updating the component designs and manufacturing processes will help considerably to establish the magnitude of these compromises. These gains are likely to be significant, perhaps because of the need to purchase fewer machines, owing to the fact that modern manufacturing equipment is more versatile, so necessitating fewer individual operations. For example, machining centres as opposed to milling machines, turning centres as opposed to centre lathes and ancillary operations etc.

Although trying to change a design or a manufacturing process is likely to cause a number of problems within an organisation, it is worth investigating any options which might exist, no matter how many eyebrows might be raised. As mentioned previously, it is essential that the temptation simply to automate the existing system is avoided. Such an approach, which alas is all too frequently adopted and not only with shop-floor automation systems, will probably lead to a far from optimal utilisation of scarce capital resources.

After having fixed the design and processing routes of the components, the next stage is to consider how many workstations will be required to enable the company to meet capacity targets, both for the short term and the long term. The desired level of automation to be fitted to these workstations has a significant influence on the number and type of workstations needed. Numerous factors need to be taken into account at this stage, for example, batch sizes, annual volumes, cycle times, level of manual intervention required for set-up operations, frequency and duration of tool changes, batch throughput time targets, ability to respond to surges in requirements etc. These issues are discussed in chapter 5.

In terms of calculation this task is relatively straightforward, albeit tedious. Once again the correct application of a spreadsheet on a microcomputer interfacing to, among other data, the component information files described earlier, will greatly simplify this process.

Having established how many workstations will be used, machine-tools, robots etc., the next stage would be to lay-out the FMS and allocate positions to the workstations and their ancillary equipment. This task should be carried out in a number of phases. First, identify any constraints which might exist. For example, what space is available and what freedom can be used in positioning equipment? Are there some columns obstructing areas? Is some of the floor area not sufficiently rigid to provide an adequate foundation for machines etc? These issues are discussed further in chapter 11.

The next phase is to consider all the feasible layouts and minimise both the number and duration of transport operations which need to be carried out. A consideration which should also be taken into account is any expected changes in product mix which might substantially affect the optimality of the layout. Possibly, the ease of relocating some equipment within the system at a later date might be relevant. This task is best carried out with the aid of a computer simulation, and is further described in chapter 7.

Then with both the type and location of the workstations fixed, some decisions as to how the workpieces should pass from one workstation to the next will have to be made. Again computer simulation will be of considerable assistance when considering the major work transport issues, for example should conveyors or an AGVS be used? The type of components will dictate some of the requirements of the material handling within the system. For example, if the parts display rotational symmetry, robots might be appropriate to move parts between closely related workstations, such as in and out of machines, while conveyors might be more appropriate to connect widely separated workstations. Conversely if the components are predominantly cubic, an AGVS might be more appropriate. These issues are discussed in chapter 6.

4.8 Concluding comments

Unfortunately one of the greatest difficulties of designing flexible manufacturing systems is the fact that decisions on factors such as those outlined above cannot usually be made in isolation. For example, additional investment in machine-tool sophistication could generate savings elsewhere within the system, perhaps by reducing the complexity of the material handling systems. Similarly, if space is at a premium one might not be able to accommodate an AGVS. Mixing rotational and cubic parts might appear desirable, but be inappropriate if the impact on fixturing is considered. A major influence will be exerted by the level of operator involvement that is intended within the system and, of course, the amount of quality control work required. If the components are too heavy, bulky or dangerous to be transferred manually, a high degree of automation will be inevitable. Alternatively, if a significant amount of operator intervention is acceptable within the system, the level of automation can be reduced.

Needless to say, a thorough analysis of the implications of all these, and any other, relevant factors is essential. The factors should be considered collectively, and then offset against each other. If eventual FMS design decisions are made incorrectly, at best the system will be both unnecessarily complex and expensive; at worst, it will not function adequately.

All these issues are intimately related to the selection of the parts to be produced. If the parts chosen lead to an undesirable design of FMS, either the sophistication of the system will need to be reassessed or a new set of parts should be selected. Most components do have designs and manufacturing processes which have evolved over many years in response to both internal and external environmental changes. Such evolution usually builds in augmented costs and inefficiencies. The possibility of an FMS manufacturing solution necessitates a rigorous and testing analysis which can lead to substantial cost and efficiency savings that will be of value regardless of whether or not the FMS is implemented.

5 The Processes

5.1 Introduction

Once the components to be produced within the FMS have been selected, many aspects of the design and manufacturing capability of the system are, almost by default, fixed. At the most superficial level, if the system is to produce rotational components, lathes and perhaps cylindrical grinding machines etc. will be required. If cubic components are to be produced, machining centres and drilling machines etc. are likely to be present. Similarly, if parts or products are to be assembled, one could expect to see assembly robots etc. in the system.

But it is not only the symmetry of the parts which dictates the characteristics of the FMS equipment, it is also features such as their physical size. If the parts are large, material handling equipment, storage areas etc. are also likely to be large, as are the cycle times on machines. Probably the mean time between the occurrence of events within the system will also be large (for example, the tape run times and the time in between set-ups etc.). Conversely, if the components are small, machines and other equipment are likely to be small, cycle times are likely to be fast, and hence the mean time between the arrival of events, such as a new set-up being required, could also be relatively small depending on the batch sizes. The purpose of this chapter is to consider how the parts selected impact on the processes which then need to be included within the FMS.

5.2 The initial analysis

Although by this stage of the project many of the more significant design constraints on the system will be defined, the major part of the detailed design which will effectively fix the overall design remains to be completed. A particularly important part of this design effort is that pertaining to the processes available to the FMS, and the workstations and ancillary equipment such as CNCs, robots, vision systems etc. which will be responsible for supporting the process workstations.

After having selected a set of parts from which a subset will be identified for manufacture within the FMS, a detailed review should be carried out of all the component designs and, where relevant, previous manufacturing processes and operation sequences. If the part is new and has never been produced before, it should be a relatively straightforward process to ensure that it is as easy to manufacture within the FMS as possible. Once again, a key issue during this phase of the analysis is to ensure that the design engineers work closely with the production engineers.

Alternatively, if the FMS is being designed for a true job-shop environment, it is likely that most of the designs of the parts to be manufactured, excluding maybe a subset of regular repeat orders, will not be known. This obviously increases both the difficulty and the risk elements associated with the design of the FMS, since it is the processing capability (for example, the component envelope and the operations to be carried out on the machine) which is being defined. If this is selected incorrectly, the FMS will not be as correctly applied as might have been possible, and it could instead not be as flexible as was anticipated, or alternatively might cost more then was really justified. To avoid pitfalls such as this, the process equipment has to be selected particularly carefully to ensure that it may be applied to as broad a set of manufacturing circumstances as possible.

5.3 Collecting the data

Having identified the parts, the next step is to establish the most appropriate processes with which to manufacture the parts, and hence their viable process routes within the FMS. At this point it is worth investigating the possibility of providing some alternative process routes, since these can be used to simplify the work scheduling task, so facilitating better equipment utilisation, and provide the opportunity to re-route parts in the event of, say, an equipment failure. Again this analysis will be much simplified if the information generated is stored within a database management system. This will greatly facilitate the sorting, optimising and presentation of the substantial amount of data involved.

To carry out this task correctly requires an immense amount of production engineering skill, together with detailed knowledge of the component characteristics and manufacturing requirements; and all this would need to be combined with a thorough understanding of how the parts are currently produced, and why. But the most important requirement is a detailed knowledge of those recently developed processes which are now available to be used within the FMS, and which are sufficiently well-proven, appropriate and able to be accommodated within the system from both technical and operator standpoints. Most production engineers will be familiar with the advances that have been made with frequently used machines such as turning centres and machining centres, especially if such equipment is already in use within the company. These can be classified more as developments within the manufacturing arena, as opposed to innovations which might represent a significant technological step, and hence conceivably risk. Possibly, advances such as these could justify an entirely different approach when establishing capabilities and ultimate requirements of the FMS. It is unlikely that the company's production engineers will be familiar with all such advances, especially with the less frequently used processes such as wire-cut, MIG and TIG welding, low melting point alloys for die manufacture, abrasive machining, phase change fixturing etc.

Indeed, it is difficult for anyone to remain abreast of technology which is developing as quickly as it is today. Unfortunately, there is no simple answer. On an individual basis, it is obviously a long-term process which should include a variety of forms of information input. The reading of trade journals, the visiting of other plants and, if possible, other countries, where the manufacturing philosophy is likely to be entirely different, are all important. Similarly, exhibitions and conferences should be attended, though the latter can be quite expensive and time-consuming, and therefore should be selected carefully.

Also, all companies tend to have their favourite suppliers with whom they have traditionally done business, especially the larger organisations. There is no reason why these should not be consulted first, but this should only be regarded as one of many information inputs, rather than the sole authority. However, factors such as maintenance are likely to be simplified if all the equipment is from one supplier. This consideration will typically be relevant in the selection of control systems for process equipment, but probably not so important for the process equipment where the demands of the parts are likely to preclude the possibility of standardisation of equipment suppliers.

If it is felt that detailed up-to-date information is not available within the company, then there is little alternative (within a short timescale) than to either engage someone who possesses the desired information or, alternatively, employ the services of a reputable manufacturing systems and process consultant, essentially as was suggested in chapter 3.

One of the major goals while carrying out the investigation into the overall manufacturing process requirements should be to minimise the number of operations needed to produce a part. An obvious and significant benefit of such an approach is that there would be a substantial reduction in the number of set-up operations which need to be carried out. Not only do these waste valuable production time, but in an environment where there are unlikely to be many operators, the risk of a batch having to wait is reduced if the number of set-ups is minimised. Similarly, by reducing the number of operations, the number of inter-workstation transfers is also minimised. This then has a beneficial effect on the computer control system etc. With production equipment continually increasing in versatility, this should be a relatively straightforward task.

Considerations such as these can result in a substantial reduction in the overall complexity of the whole manufacturing system. Of course, it is likely that the process equipment selected will be proportionally more sophisticated as a result, but one should not be deterred from the essential need to reduce to an absolute minimum the number of operations required to turn the raw material into a finished product. It is not unusual for the number of operations required to produce a part to be reduced to a quarter of what was previously needed, and sometimes even greater reductions can be obtained.

As an aside, it should be noted that this type of analysis is likely to prove equally beneficial to a company which is only investing in say CNC, as opposed to full FMS. Many of the advantages described above would be available to a CNC cell. If planned carefully, such an investment could quite reasonably be regarded as the first step towards a full FMS. It is a well-documented fact that many of the most easily quantifiable benefits from installing an FMS are in fact generated by the type of analysis described above — that is, by imposing the disciplines necessary to implement an FMS rather than by actually implementing the FMS. Without doubt, there is much to be gained by approaching FMS design in a well-planned and disciplined manner. The eventual gains from full FMS are immense, but that does not mean they all have to be reaped at the start of an automation programme.

5.4   Equipment requirements

Having carried out a survey into the production processes and equipment currently available, and having selected the most appropriate options, the next step is to decide how the parts will be produced on the equipment and how many process workstations will be needed to accommodate the required volume of parts. At this stage, due consideration should be given to both the short-term and long-term manufacturing goals (possibly by leaving space for other equipment to be added later), and other factors, such as the ability to handle surges in demand.

To establish how many of the various types of workstations will be required within the FMS, a table should be compiled containing the expected process times for each part being considered for production by the FMS. An example of how such a table might be prepared was shown in figure 4.2. However, such a table would now need to be enhanced to reflect the increased level of knowledge concerning the part, not only its current process routes but also the possibilities offered by employing new manufacturing techniques. This facilitates the comparison of the old and new techniques, and a quantification of the gains likely to be made in terms of reduced inventory and reduced manufacturing time.

During this analysis, the influence of a variety of high-level factors will need to be taken into account, for example:

- Planned maintenance
- Unplanned maintenance
- Unavoidable idle time
- Set-up times
- Load and unload times
- Tool changing times
- Inspection times
- Transportation times
- Operator imposed idle time
- Tape run times

Having established how much time will be needed on each process workstation for the parts being considered, it is then necessary to calculate how much manufacturing time will usually be available to the plant. Other factors which would typically need to be taken into account would be:

- Working hours per shift
- Shifts per day
- Working days per week
- Working weeks per year

- Hence, working hours per year (A)

- Part numbers
- Parts needed per year

- Annual hours required
- Allowance for planned maintenance
- Allowance for unplanned maintenance
- Allowance for unplanned idle time

- Hence, total hours required per year (B)

Then dividing the number of hours required (B) by the number of hours available (A) will give the expected number of workstations required, with fractions which generally will have to be rounded up. An example of how calculations such as these might be presented is shown in figure 5.1, as usual with the summary sheet being supported by a number of cross-referenced related data sheets.

Sheets such as these are simple and straightforward to prepare, yet they provide a clear and concise description of the logic used to calculate, in this case, the number of workstations required. All such calculations relating to system design parameters should be well documented, including any relevant assumptions, which should be in readily accessible locations to which all members of the project team have access. Since it is likely that much of the

ALL–PARTS–MADE–FAST LTD,
Plant 3 — Flexible Manufacturing System Project

CAPACITY DATA	Title: CD1	Ref to: CD/R1	Date:
			Author:

Name: CNC turning centre 1 (1988)   Sheet .... of ....

Part No.	Op. No.	Cycle (min.)	Parts/ annum	Idle hours	Hours needed	% time working	Comments
A123456	20	1.5	300	3.0	13.0	0.35	Rough turn
	60	2.0	300	4.5	17.0	0.46	Finish turn
D234352	40	1.9	100	2.2	6.2	0.17	R/turn + drill
D324548	30	2.1	250	3.8	14.6	0.40	Rough turn
N3594	30	1.1	400	4.1	14.8	0.40	Rough turn
	70	2.8	400	5.3	27.3	0.74	F/turn + drill
N8542	20	0.8	100	1.9	4.1	0.11	Rough turn
S457534	20	1.3	50	2.1	3.6	0.10	Rough turn
etc.							
				26.9	100.6	2.73	TOTAL

ALL–PARTS–MADE–FAST LTD,
Plant 3 — Flexible Manufacturing System Project

RELATED DATA	Ref: CD/R1	Date:
		Author:

Name: CNC turning centre 1 (1988)   Sheet ...... of ......

Comments: ....

....

Assumptions:
1 shift/day x 16 working hours/day x 5 days/week
= 80 working hours per week
46 working weeks/year = 3680 working hours/year
Set–ups allowed at .... (tool/jig availability)
Operator relaxation allowance at ....
Maintenance (planned and unplanned) at ....
Allowance for scrap included at ....
Allowance for spares included at ....

Conclusions:
Number of workstations required is calculated by dividing % time
    working by 100 and rounding up to nearest integer
    etc. ....

Figure 5.1   An example of a typical capacity data sheet

equipment will be new, many of the processes could also be new. This implies that many of the cycle times will have to be estimated, in which case the analysis is bound not to be entirely accurate. In this case, it is particularly important that any assumptions made during the calculations are clearly stated.

Having developed these tables, it is then appropriate to study the conclusions and check that the parts selected are indeed ideal for manufacture within the FMS. For example, care must be taken to ensure that certain workstations are not severely under-utilised, perhaps only being required for one or two parts, and that the raw materials required to be within the system are only of a few types, thus simplifying swarf or chip removal, or in an assembly environment ensuring that no trivial and yet awkward items have to be provided for a very small subset of the parts to be produced. Again, this is likely to be an iterative procedure, with a variety of different components and processes being considered. After this analysis has been carried out, and conclusions evolved, the set of parts selected for manufacture within the FMS should, after one further check, be fixed, and only changed if serious unforeseen circumstances arise.

5.5 Checking the analysis

With the set of parts selected, it is highly desirable that the somewhat 'mechanical' analysis procedure described above is subjected to some testing, largely to confirm the validity of the assumptions made. The only fast and efficient way to achieve this is by using a computer simulation. Even though simulation is discussed in considerable detail in chapter 7, it is nevertheless appropriate to make a few comments at this point.

Without doubt, computer simulation is one of the most valuable tools available to the FMS designer. This is particularly the case now that modern microcomputers are sufficiently powerful to run a sophisticated modelling system. At an early stage in the development of the FMS it is well worthwhile starting to use an appropriate modelling package to confirm the facility layout, material handling routes and, in particular, the number of workstations required.

To ensure the analysis is comprehensive, a number of 'runs' of the model would be required. The basic inputs would be a variety of different work schedules for an appropriately significant time period (say between three and six months). Although the actual layout of the FMS and the selection of the material handling systems are unlikely to have been fixed, the more attractive options will probably have been identified. As many as possible of these should be used. Of course, the work schedules should include best-case and worst-case combinations of different batches of parts in order to simulate the effect of blockages and imbalances within the FMS. The effects of planned and unplanned maintenance should also be included.

Obviously, this analysis could represent a substantial amount of work. If, for example, there are 10 different four-month work schedules, 3 layout philosophies and 5 different workstation combination options, then 150 simulation runs would be required. It is therefore important that the modelling system selected permits the 'batching' of runs so that the minimum operator input is required to actually carry out all the simulations, and that the results may be displayed and interpreted easily. An example of how such results could be presented is shown in figure 5.2, where the utilisation of various facilities (machine 1, robot 1, storages 1 and 2 etc.) are tabulated for a number of schedules as two layout options are considered. The end result of this work will be to ensure that a greater degree of confidence exists in the set of parts selected as being appropriate for the FMS.

The type of features which would typically need to be considered are essentially as follows:

- Load and unload bays used
- Setters, operators and maintenance engineers used
- Process workstations used
- Material handling required
- Storage required
- Inspection stations required

5.6   Equipment utilisation

As result of these analyses, one will be able to commit to a specific number and type of workstation to manufacture the components selected. Obviously, the more costly the workstation, the more carefully one would wish to investigate the conclusions of these analyses. This is especially so, if, as is likely, the components are to be passed from one workstation to another, with a fairly regular trend being recognisable (that is, some workstations for roughing operations and others for finishing), since as parts progress to later operation equipment, it becomes more difficult to maintain the high level of utilisation which would be expected of earlier operation equipment. This is due to the effect of any problems and/or inefficiencies in the earlier stages of production being amplified by the time the parts reach the later stages. Therefore, it is prudent to ensure that the more capital-intensive equipment is, wherever possible, used for early operations within the system. Otherwise, the inevitable inefficiencies in, say, the material handling system or inability to carry out a set-up on time will result in unnecessarily expensive idle time being imposed on these workstations.

In fact, as a general rule, the system should be designed such that, even allowing for the worst possible combination of circumstances in the earlier

ALL–PARTS–MADE–FAST LTD,
Plant 3 – Flexible Manufacturing System Project

| SIMULATION DATA | Title: SD1 | Ref to: SD/R1 | Date: |
| | | | Author: |

Purpose: Layout v. schedule .... Sheet .... of ....

Sch. No.	LAYOUT 1							LAYOUT 2							Comments
	M/C1	RBT1	STR1	STR2	etc.	FIN	WIP	M/C1	RBT1	STR1	STR2	etc.	FIN	WIP	
1	66	21	5	8		214	26	71	24	6	8		205	25	
2	43	15	3	4		289	20	47	19	2	5		281	18	
3	72	24	6	10		178	31	79	28	7	9		166	26	
4	85	28	7	12		162	33	88	32	8	9		153	32	
5	29	8	1	1		381	12	34	15	3	4		362	11	
6	58	19	4	6		202	19	61	26	2	7		189	16	
7	64	22	5	8		218	23	68	25	6	8		208	20	
8	79	26	6	10		170	27	82	33	4	5		162	23	
9	82	27	7	12		163	22	86	31	3	6		150	18	
10	69	24	5	8		179	25	75	29	5	7		176	22	
	64.7	21.4	49.0	79.0		216	23.8	69.1	26.2	4.6	6.8		205	21.1	Average

ALL–PARTS–MADE–FAST LTD,
Plant 3 – Flexible Manufacturing System Project

| RELATED DATA | Ref: SD/R1 | Date: |
| | | Author: |

Name: .... Simulation Data .... Sheet ...... of ......

Comments: .... Layout v. Schedule ....

....

Units:
Workstation utilisation expressed as %
Storage utilisation expressed as average contents
Finish time expressed in days to complete schedule
WIP expressed as cumulative system average

Assumptions:
Layout 1 has no return loops on AGV paths; Layout 2 does
Schedules 1,2,3 etc. represent ....
Maintenance and breakdowns assumed at ....

Conclusions:
etc. ....

Figure 5.2   An example of a computer simulation data sheet

workstations, the later workstations will always be able to handle the work generated. In these circumstances, it is highly unlikely that bottlenecks will be created. This makes it much easier to maintain high utilisation and output levels within the entire system. Then, if the first operation workstations are fully utilised, the whole FMS is approaching full utilisation.

Having established a means of determining what is likely to represent a fully utilised state within the FMS (albeit for a certain set of parts), one then has to decide whether or not this level is acceptable, both as an average for the system and on an individual workstation basis. It is difficult to generalise as to what is an acceptable level of workstation utilisation (and indeed many people have widely differing opinions as to the definition of utilisation). However, acceptable figures for actual working time (for example, a machine-tool in operation) could be anything between 70 and 80 per cent. However, in certain circumstances, say for either a very important, or at the other extreme, for a very unimportant workstation, one could expect the quoted bounds for these figures to be exceeded substantially. However, one should take into account such factors as:

(1) Although the overall utilisation of a machining centre might be as high as 75 or 80 per cent, the proportion of time it is actually cutting metal could be only 40 per cent (which, in fact, is not an unrespectable time).
(2) If the workstation is a robot which has purely to load and unload, say a turning centre, its overall utilisation might be as low as 10 or 20 per cent. The justification for this could be that it would not be acceptable to have the machine waiting for the robot to carry out a load/unload cycle, even if the robot was shared between two or more workstations.

Another important part of this analysis, and indeed all such analyses which are carried out as part of the FMS design procedure, is a quantification of the accuracy and sensitivity of the results obtained so far. This entails carrying out yet another analysis. During this, significant influences on performance should be varied and their impact on features, such as workstation and operator utilisation, monitored. Examples of influences which should be considered are surge capacity, an increase in the maintenance needs of an important workstation, an increase in estimated cycle times etc. The overall effect on the productive capacity of the entire system can then be identified, and an appreciation gained for the effect of one being, say 10 per cent out on some of the estimated calculations.

If this analysis is performed with the aid of a computer simulation, then the task will be relatively straightforward, being essentially as outlined previously. However, if manual techniques are to be employed, then care should be taken to allow sufficient time to carry out the analysis comprehensively and to display the results clearly.

Factors which would typically need to be taken into account include:

- Expected availabilities
- Utilisation targets
- Nominal (expected) requirement
- Best-case and worst-case load requirements

An example of how such an analysis might be presented is shown in figure 5.3, where a number of parts are listed together with their expected, best and worst annual volumes and production hours' requirements.

5.7 Ancillary equipment

Having established the type and number of items of major equipment to be included within the FMS, the next step is to decide what types of ancillary equipment will be needed to support the activities of the process work-stations. Obviously, the equipment and features to be considered are somewhat application-dependent, but some of the options which would typically be considered are as follows:

- NC, CNC or non-CNC?
- Should communications be MAP compatible?
- Colour graphics at the CNC console?
- In-process gauging equipment?
- Closed loop machining?
- Post-process gauging equipment?
- Quick change tooling?
- Tool carousels (and their capacity)?
- Tool wear compensation?
- Tool breakage detection equipment?
- Quick change fixtures?
- Component turnround fixtures?
- Combination jaws?
- Automatic pallet shuffle mechanisms?
- Standardisation of sensors?
- Standardisation of plugs, sockets and cables?
- Automatic vision inspection equipment (parts and/or tools etc.)?
- Automated swarf removal?
- Level of workstation diagnostics?
- Closed circuit TV monitoring systems?

These options and many more, will need to be considered in the light of such factors as the planned level of automation and hence operator presence

ALL—PARTS—MADE—FAST LTD,
Plant 3 — Flexible Manufacturing System Project

SENSITIVITY ANALYSIS	Title: SA1	Ref to: SA/R1	Date: Author:

Name: CNC turning centre 1 (1988)   Sheet .... of ....

Part No.	Op. No.	Prts /yr			Hrs reqd			% Prob.		Comments
		Wst	Expt	Bst	Wst	Expt	Bst	Wst	Bst	
A123456	20	250	300	400	11.3	13.0	16.3	5	10	
	60	250	300	400	14.9	17.0	21.2	5	10	
D234352	40	100	100	200	6.2	6.2	10.2	10	5	
D324548	30	150	250	450	10.3	14.6	23.3	5	25	
N3594	30	350	400	500	13.4	14.8	17.4	5	10	
	70	350	400	500	24.6	27.3	31.6	5	10	
N8542	20	0	100	150	0	4.1	51.5	0	5	
S457534	20	0	50	100	0	3.6	5.1	0	5	
etc.										
		1450	1900	2700	80.7	100.6	176.6			TOTAL

ALL—PARTS—MADE—FAST LTD,
Plant 3 — Flexible Manufacturing System Project

RELATED DATA	Ref: SA/R1	Date: Author:

Name: CNC turning centre 1 (1988)   Sheet ...... of ......

Comments: ....

....

Assumptions:
Best—case volumes imply ....
Worst—case volumes imply ....
Expected volumes are calculated on the basis of the
    assumptions shown in Related Data sheets ....
Nominal loads calculated on the basis of ....
Worst—case loads calculated on the basis of ....
Expected utilisation target ....
Parts/year calculated on the basis of ....
Hours reqd calculated on the basis of ....
Probability values calculated on the basis of ....

Conclusions:
    etc. ....

Figure 5.3   An example of a sensitivity analysis sheet

within the FMS, funding availability etc. Obviously, if there are to be virtually no operators in the FMS, a greater degree of intelligence is likely to be required at the workstations. Another consideration would be the long-term requirements of the system. Is this system an end in itself, or is it merely one of a number of automation steps that the company intends to take?

Frequently, it is assumed that most equipment within an FMS will be of a comparable level of sophistication, in automation terms. (This is probably a result of the fact that most of the earlier systems were implemented on green-field sites by individual machine-tool vendors who were keen to use as much of their own new equipment as possible.) Suffice to say, this comparable level of sophistication is definitely not a fundamental requirement of successful FMS. Indeed, there are considerable advantages to be gained in certain circumstances by carefully mixing the technology levels within a system, not least by reducing the capital investment needed, since possibly not all the equipment needs to be new or so highly sophisticated.

Probably the most significant example is when an FMS is designed to include both CNC and non-CNC machine-tools. Many people feel that such a combination of equipment cannot be made viable within a true FMS. This is not the case, as has been admirably demonstrated in several systems, such as in SCAMP. Indeed, it has been suggested that because non-CNC machines are simpler in control terms — that is, they are either working or they are not — they are, in fact, easier to integrate into an FMS environment. Of course, against this has to be weighed the fact that they are likely to be more difficult to set-up, and less versatile.

Although much of the ancillary equipment to be selected will be fundamental to the integration of the workstation into the FMS, certain equipment is likely to represent an optional 'luxury'. The incorporation of these items could depend on a number of the system's basic objectives, such as:

- How long must each workstation be capable of operating without manual intervention?
- Are part programs to be prepared and modified at the workstation controller?
- How wide a family of components is the system to be able to accommodate?
- What level of automation/demanning is expected within the system?
- What level of process control is required?
- What technology can easily be accommodated by the organisation?
- What financial constraints exist?
- Is the system likely to be extended later?

The situation might well be confused slightly by the fact that much of the equipment will need to be designed or significantly modified to suit the requirements of the specific FMS. As such, it might not yet be clear which level of automation is required to, for example, ensure that the set-up operations are as short as is desirable, or the extent of automation which was assumed in the workstation capacity and sensitivity analyses.

As will be the case with a number of the remaining equipment and system functionality issues, the desirability of incorporating features such as colour graphics or a sophisticated part program editing facility at a CNC control (or indeed even at the FMS control consoles) is likely to depend on the programming philosophy of the system. If part programs are rarely prepared or changed, there is little point in providing a sophisticated editing facility at the CNC console (though admittedly good colour graphics, for example, do significantly improve the quality of the user interface of most controls, regardless of whether or not programming is being frequently carried out). Indeed, if the ability to modify programs is readily available at a workstation, great care must be taken to ensure that the correct version of the program is both stored on the host computer system and used, if appropriate, with the next batch of parts.

However, if the FMS is to be installed within an essentially job-shop environment where parts are likely to be produced in very low quantities, perhaps only one, and then not be required again for a lengthy period of time if ever, then it might well be considered desirable to provide a part programming facility at the CNC control for the use of appropriately skilled personnel. In these circumstances, such a system might, for example, include tool path graphics, and automatic depth of cut and spindle speed selection, together with material clearance routines or 'canned' drilling cycles to help facilitate the generation of the program, and improve the likelihood of it being correct first time etc. To avoid imposing unnecessary idle time on the machine, it should be possible to create a new part program while an existing program is being executed.

Certainly it is difficult to generalise on most of these issues. For example, concerning CNC or non-CNC machines, one of the main problems is likely to be the importance, duration and complexity of the manufacturing operation to be carried out. It is unlikely that a non-CNC machine would be used for a complex milling or turning task. On the other hand, there are some operations for which CNC machines are not yet, or only very recently, becoming available, for example, spline milling or hobbing, large presses and broaching machines. So, generally one would only expect to find non-CNC machines as later operation machines where larger set-up times and lower utilisations are unlikely to cause a bottleneck or accusations of poor capital allocation. Indeed, since non-CNC machines tend to be less costly then their CNC counterparts, the incorporation of such machines as later operation machines within an FMS is often a distinct advantage. Considerations such as this can be used to simplify the choices to be made regarding equipment selection.

Concerning the ancillary equipment, it is desirable that all the workstations are as similar in operational terms as possible in this respect, for example, by having all the workstations automatically loaded and unloaded perhaps by robots with the same program architecture. This may not always be possible, but where it is, considerable system simplification (in terms of design and operation) can result. If such an approach does not prove practical, then care should be taken to ensure that manual dependency is not built into the system from the outset, unless an operator is to be present at the equipment all the time. Otherwise these are likely to become bottlenecks, since system performance will be operator-dependent, and hence not as predictable as the rest of the system since the presence of an operator could not always be guaranteed. If such an approach has to be employed, the user of the FMS should be aware of the possible shortcomings of such a configuration and allow appropriate capacity contingency plans for the workstations concerned.

This analysis would continue until all the relevant ancillary equipment options have been considered, their costs, benefits and drawbacks quantified, and a decision made as to whether or not the feature should be included within the FMS in either the short term or the long term.

Although the above statements are intended to be as generic as possible, they will need to be interpreted in terms of the particular manufacturing environment in which the FMS is to be installed. For example, while the issue of having excess capacity in the later operation workstations is an extremely important one in virtually all manufacturing systems, machining, process, assembly or fabrication, the issue of CNC *versus* non-CNC machines might well not be applicable, not least since all the equipment might have to be virtually the same, perhaps assembly robots etc. Ultimately, within flexible manufacturing systems, it is important to provide an appropriate balance between capacity and flexibility.

5.8 Inspection equipment

One of the most significant advantages that an FMS offers over traditional plant is the opportunity to produce components of consistent accuracy and quality. The selection of appropriate inspection equipment and the definition of clear quality control disciplines is essential if these benefits are to be achieved. Although this subject probably merits a book in itself, this section is intended to provide a sufficient insight to the major principles to ensure a potential system designer does not overlook the key issues.

The selection of inspection and gauging equipment will largely depend on the needs for accuracy of the components and whether or not the batch sizes are greater than one. There are four approaches which need to be considered, some of which may not be relevant to a particular application, for either technical or other reasons:

- Manual inspection
- Automatic inspection
- In-process
- Post-process

If the FMS is to machine valuable components, perhaps a turbine disk from an aircraft engine, in-process gauging would be highly desirable for some operations. Such a system might well be based on probes, which are becoming increasingly popular. Selection of these would essentially enable the machine to become a gauging machine, within certain limitations. There are some advantages associated with such an approach. Since the part is still on the machine it might be possible to carry out a cycle of gauging followed by further machining, and hence slowly and carefully reduce the size of the part to the required dimension. This should minimise the probability of the part being over-machined, though usually while the machine is busy gauging it is not carrying out the prime task of removing material. But, because machines are designed for metal removal, with specific production processes in mind (one of which is not gauging) it does not necessarily follow that they will be particularly accurate for gauging. Dimensional changes due to expansion, contraction and vibration of the machine could well make a nonsense of in-process gauging readings.

However, if the part is gauged after the process is complete, there might be other problems. Certainly if the part has been incorrectly machined it is not likely to be easily rectified. If insufficient material has been removed, it is not possible simply to send it back to the machine and make additional cuts. If too much material was removed, not only might it be impossible to rectify the part but, in the meantime, the next part, or parts, might also have been scrapped. In this event, if one is sampling components post-process, one might well have to ensure that the software controlling the entire system has the capability of tracking the parts produced during the period when the relevant part found its way to the post-process gauging station and the part currently in produced reached the machine. Such software is definitely not straightforward, as is discussed in chapter 8.

However, one distinct advantage of post-process gauging is that at least it can be carried out on the most appropriate equipment, in the best possible environment. Certainly, the advances that have been made with co-ordinate measuring machines have been substantial, making them considerably easy to integrate into an FMS installation.

As with so many aspects of FMS design, in-process *versus* post-process gauging is a compromise. The relevant cost and benefits will have to be considered in the light of the requirements and constraints of the particular system. Either way, it should be emphasised that ultimately it is the process which should be controlled. If the process is functioning within tolerance, then the parts produced will be within tolerance. In this case gauging becomes checking, rather than establishing what has been produced and how the process should be modified to ensure better parts are produced in the future. Any costly gauging procedures should be restricted to the strategic work-stations rather than the less-important operations, which could well be gauged later, collectively.

5.9   Control of the process

To some degree, virtually all gauging is post-process. Only a few systems exist which carry out true in-process inspection, an example of one being the cylindrical grinding machine used within SCAMP, which was able to gauge a part during the grinding process, so ensuring that it was within tolerance.

It is interesting to note that much of the success of many Japanese manufacturing systems frequently attributed to philosophies such as 'Just in Time' etc. is, in fact, due more to commitment and good process control, which results in both predictable work flow and system performance. For example, if the process is accurately controlled, inspection requirements become less stringent since one has a greater degree of confidence in the part already being correct.

Control of the process naturally requires one to consider incorporating such features as tool wear compensation, tool breakage detection, spindle power monitors etc. The reliability and consistency of all these options have been improved considerably over the past few years and are now useful additions to sophisticated production equipment.

Tool wear compensation systems frequently use probes to measure the cutting dimensions of the tools, with new offsets being input directly into the control system of the machine. The probes may also be used for checking the presence or absence of the tool, which, for example, with a drill, might be a means of checking whether or not the tool broke during the preceding cutting operation.

Spindle power monitors have not had such a rapid rise in popularity as probes, largely because of the difficulty in correlating the relatively 'noisy' readings with the relatively small trends developed as a result of the wearing of the cutting tool (that is, low signal-to-noise ratio).

5.10   Tooling and fixturing

Tools carousels and automatic tool change equipment are an important part of nearly all flexible manufacturing systems, regardless of the discipline, machining, assembly, fabrication etc. These are present usually as a result of the desire to operate the equipment for long periods without manual intervention. The speed of automatic tool changers is usually dependent on the mechanism employed, but the capacity of the tool carousel is best decided with the aid of computer simulation, unless, as in some systems, it is intended to exchange the entire contents of the carousel at the start of each batch. Generally, carousel capacity is dependent on the number of tools required to complete the most complex part, since, if possible, one does not want to have to change tools in mid-component, and ideally not even at mid-batch. Within a good FMS it is always obvious that care has been taken to ensure that operator interactions are (1) kept to a minimum and (2) all occur simultaneously at a particular workstation. This makes it much easier for scarce resources such as operators to schedule their own availability.

The selection of pre-set quick-change tooling and the precise form of a tool carousel will be application-dependent. But in environments where manual intervention is inevitable, every effort should be made to ensure that complexity and duration are at an absolute minimum.

Similarly, an appropriate combination of pallet shufflers, gantry loaders and robots (all of which will be discussed in detail in the following chapter) is likely to be required if the system is to operate for any length of time without manual intervention. Many of these are now included within machine-tool specifications, purely as optional extras. Again the benefits of increased automation will have to be weighed against factors such as increased cost, operator skill requirements, maintenance and system complexity etc.

The issue of work-holding continues to represent a significant problem for all highly automated systems and, like that of inspection, probably merits a book in itself. Suffice to say, it remains extremely difficult to fixture a wide variety of parts with a small number of jigs and fixtures. Frequently, pallets for a machining centre to be located within an FMS will be quite expensive. The fixturing equipment to go on to the pallets is likely to be at least as costly.

With cubic components of complex geometries, the currently preferred approach to fixturing appears to be with 'kits' of standardised fixture sets which then need to be assembled in a particular way to hold the part in question. The ease with which this can be done might well be increased significantly when solid modellers are used more frequently to create both the part programs and details of the fixturing required to hold the parts. Some so-called 'universal' fixturing systems have been developed (see chapter 12). However, none of these has proved to be truly universal and hence has not been received as a generally accepted standard.

For turned or rotational components, to some degree the problem is slightly simpler because of the guaranteed symmetry of the part. Again though, the possibilities for simplifying the situation with the use of solid modelling systems are substantial. However, with more sophisticated chucking devices becoming available, equipped with automatic jaw changers etc., the problem for most rotational part FMS is not quite so severe, though decisions such as whether to use hard, soft or combination jaws still remain.

One of the most significant differences between the work-holding of turned components as opposed to cubic parts is that usually turned components are processed in different fixtures to those in which they are transported. This is particularly the case with small components, though perhaps with the very large parts both chucks and components might well be transported together. The symmetry of these parts is again useful since their centreline may be located quite accurately and easily with the aid of relatively simple devices, such as 'vee blocks'. There could be one of these at each end of the part, providing a simple and inexpensive means of transporting the component horizontally. Alternatively, one 'vee' and some form of 'cup' might be used if the part is being transferred almost vertically. Depending on the stability and dimensional variations of these parts, a flat surface might be all that is required to transport the part in a known orientation. Examples of these approaches are shown in figure 5.4. An important added advantage of, for example, 'vee blocks', is that they can also be used as a mid-cycle turn-round fixture.

Some of these techniques could also be used to hold cubic components, as is shown in figure 5.5. Of course, one of the issues to be remembered is that once the part has arrived at the workstation, it then has to be passed to the process device via a material handling device. The work-holding fixture used for transport should be well suited to the material handling device in particular. It could well be that an optimum design feature for one part of the system is not well suited to the activities within another part of the system.

Although the fundamental aim of all fixturing is to locate components accurately during processing (usually with minimum investment) it should also be noted that the efficiency of location is also likely to have a significant impact on the duration of set-up operations.

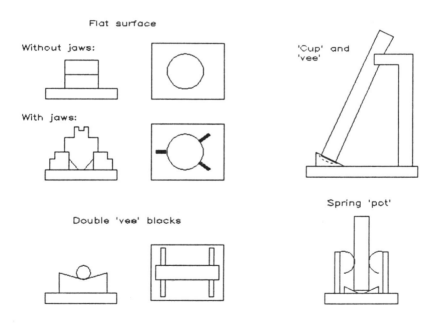

Figure 5.4 Examples of how rotational components may be located

Figure 5.5   Standardised fixturing for rotational and cubic components
(courtesy of Werner Kolb)

5.11   Set-up operations

Because components are being produced in relatively small batches, perhaps as low as one, it is imperative that the duration of the set-up operation is as short as possible, namely the time needed to change a workstation from producing one part to being able to produce another. Obviously, any idle time is expensive, since not only is the workstation not carrying out the task for which it was designed and installed but also the set-up operations, such as manual inspections, are likely to require the presence of a scarce resource, namely a skilled operator.

Set-up operations, if at all possible, should be scheduled to occur at the most convenient time (when all the resources required are readily available), and should be of as short a duration as possible. The FMS control system should be able to help with this to some degree by, for example, checking that the required resources are all available at the required location — perhaps by sending the workstation set-up details (that is, instructions to assist the operator) and the new part programs immediately the last part of the previous batch has been completed etc. Also the FMS control system should send the operators information about the set-up, say, 15 to 20 minutes before the operation is actually scheduled to occur. This will give the operators the opportunity to prepare for the work, gathering tools, jigs and fixtures etc. and moving to the correct location to guarantee that the set-up will be carried out as quickly as is reasonably possible. Conceivably some of these activities will be automated, for example, fixturing might arrive with the part on a pallet, or tooling might be changed by automatic drum exchanges etc. However, if these activities are manual they must be made to execute as simply and quickly as possible without sacrificing the consistent accuracy desired of the task.

To put the issue of set-ups in perspective, it is the duration of set-up times which has forced, for example, many motor car and machine-tool, and indeed many other, industries who have invested in automatic machines, to produce large batches of parts even if only a small quantity is required. It is the restructuring of these traditional markets which has led to the need to manufacture in small batches, though obviously the need to produce economically remains. Therefore both the duration and frequency of set-up operations must be reduced to an absolute minimum wherever possible. While the correct selection of workstations is of fundamental importance to the success and longevity of the FMS, so is the selection of the correct ancillary equipment.

5.12 Concluding comments

Attention to detail during the design of the FMS is fundamental to the success of the system, even when, for example, such relatively small items as sensors are being considered. Although significant advances on sensor technology have been made during the past few years, the tasks which have been asked of sensors have become significantly more complex. Many of the earlier FMS facilities were deemed to have failed because it was virtually impossible to maintain a high level of productivity from the system. Frequently this was not due to any fundamental design fault within the system, but purely because too much reliance was put on a sensor input which, if it failed, was difficult to replace.

Sensors are available in a variety of forms, from sophisticated vision systems to limit switches and proximity sensors. Limit switches tend to be somewhat unreliable and, as a general rule, proximity sensors are preferable. These are non-contact devices with no moving parts. When either is used, care should be taken to ensure that it is easy to replace, perhaps by using high-quality plugs and sockets rather than wiring the cable directly into a terminal box. Additional installation cost will be substantially outweighed by reduced system downtime and maintenance costs. Also, the operation of the somewhat expensive system should not be made dependent on the input of a sensor which probably costs a relatively insignificant amount when compared with the entire FMS. If the need does arise, two sensors should be used, then if one fails the system continues to operate. In any case, it should be possible to replace sensors while the FMS remains operational.

Eventually, it is likely that the FMS will be required to operate in a 'demanned' environment, and it is essential that smooth and consistent operation of the system as a whole is guaranteed. For this to occur, a number of factors must be taken into account:

- The system should not be overstressed.
- A high level of maintenance diagnosis should be available.
- The environment should be inherently safe.

The system could be overstressed in a number of ways, all of which will eventually lead to an unnecessary drop in production. For example, tooling and the workstations could be run at maximum speeds instead of a safe fast speed. This could lead to premature tool failure, resulting in time wasted while the tool was changed, perhaps manually (and an operator may not be available), and even the component in the workstation might be scrapped. Additional maintenance might be required by the workstations because the system is being run too fast. Demands for flexibility outside the designer's intentions might be made, for example by introducing new materials for which the workstations were not intended. All these influences and many others will almost certainly lead to the system failing more often than it should. Once an FMS has failed, in part or as a whole, the problem of restarting is often significantly more difficult than within a traditional manufacturing environment. However, equivalent problems within an FMS are usually significantly more viable.

An important part of ensuring the FMS remains productive is gathering comprehensive diagnostics. This includes a variety of factors from tool wear statistics, through coolant and lubricant requirements, to, for instance, vibration characteristics of main spindle bearings. Much of this data is not easily obtained, but its collection remains extremely valuable. Since one of the secondary aims of FMS should be to remove paperwork from the factory floor, this data should be analysed by the controlling computer and initially only summary trend analysis data made available. If the data collected is adequate, and the means of analysis appropriate, it should be very straight-forward to identify factors which are having an adverse influence on the consistency and reliability of the output of the system. Naturally, quality control data is a useful addition to this information.

Finally, and probably most important of all, the processes within the FMS environment must be designed to be inherently safe. This does not only mean that obvious areas of potential danger should be adequately guarded, such as robots, rotating spindles, sharp tools etc, but also that machines do not spread swarf, coolant or hydraulic oil over the floor.

All of these factors, which once again are only commonsense manufacturing engineering, if addressed adequately will help to ensure that the process equipment within the FMS continues to carry out the production task for which it was intended, reliably and efficiently, for a long period of time.

6 Material Handling

6.1 Introduction

One of the prime motivations for developing an FMS is to ensure that the transformation of raw material to finished parts is as rapid, efficient and controlled. For this to be the case, much certainly depends on the inherent efficiency and reliability of the various manufacturing processes. But as has already been discussed, since only a relatively small proportion of component throughput time is actually devoted to processing, and also given that it is unlikely to prove possible to increase the overall speed of processing significantly, it is really the efficiency of the material handling systems which dictates the overall efficiency of the FMS. Indeed, it has been suggested that approximately one-third of a product's total manufacturing cost is absorbed by the expense of successive material handling tasks (though this does include final distribution costs).

So, following the selection of the components to be produced by the FMS and the processes to be used to manufacture the parts, the next fundamentally important area to be addressed is that of the material handling system, or probably systems. These represent the way in which parts and possibly tools etc. are transported throughout the FMS, both between integrated subsystems of the FMS (frequently called *bays*) and within the subsystems themselves.

6.2 Types of material handling needs

Usually, within an FMS there are a variety of material handling tasks to be performed. Some of these are virtually fixed, since they arrive as an integral part of a particular piece of process equipment. Selection of the others is almost entirely at the discretion of the FMS designer. In all, there are four major types of material transport tasks which need to be addressed:

(1) Transport between different systems.
(2) Transport between different subsystems within the same system.

(3) Transfers between the workstations comprising the various subsystems.
(4) Transfers within the workstations themselves.

Although it might appear desirable to address all these categories simultaneously with the same solution, it is unlikely to be possible within most systems, because of the wide varying nature of the tasks concerned. So, there is likely to be a mixture of material handling systems within any one FMS. This is an unfortunate though realistic state of affairs since it implies that not only must the selection of each material handling system be optimised in isolation but also collectively with the other systems selected.

An example of such a situation would be within an FMS designed to produce small turned components. Transfer of parts between the workstations within a certain bay or subsystem might be handled admirably by a robot. However, within an FMS producing larger components, this task might be more appropriately handled by an overhead gantry system, perhaps also because of the different geometry of the parts, machine layout or different cycle times etc. In either case, transport of the parts between the FMS and perhaps its Automated Storage and Retrieval System (ASRS), or perhaps between different flexible manufacturing systems, might be carried out using an Automated Guided Vehicle System (AGVS). This scenario has already created the need for several material handling systems within the one factory, and numerous problems are likely to be faced when interfacing all these systems to one other in both mechanical and electronic terms. To compound these problems, it should be appreciated that not all the material transport systems will necessarily be automatic. It could well be that automatic, semi-automatic and manual systems all have to be able to be interfaced as well.

The material handling systems are likely to have to carry many loads, apart from just parts. Indeed, some of these tasks could be more demanding, perhaps in terms of speed of delivery requirements compared with those of workpiece transfer. Or conceivably, the components could be awkward to handle, because they are very large or very small, or possibly because they are hazardous, in which case rigorous safety requirements might considerably complicate the material handling task.

As is the case with most aspects of FMS design, a number of sometimes contradictory factors will have to be taken into account collectively before a decision can be made as to the selection of any particular material handling system. Stated below are the main types of loads which many if not all the systems will have to carry:

- Workpieces
- Tooling
- Jigs and fixtures
- Coolant and lubricant

- Swarf
- Miscellaneous equipment
- People (?)

The next task is to identify the factors which are of particular importance and hence need to be optimised (or at least, near optimised) when selecting the material handling systems. This analysis will to a large degree be application-dependent; the main factors which typically will be of relevance are as follows:

- Speed at which transfers must occur
- Frequency at which transfers must occur
- Volume and weight requirements of the load
- Accuracy required of component location
- Accuracy required of transport system docking
- Number of elements comprising the load
- Load fixturing requirements
- Routeing flexibility and extendability
- System permanency
- Safety requirements, guarding implications, maintenance

The types of material handling system most often found within an FMS usually belong to one of the six categories shown in figure 6.1. However, it should be noted that although all material handling systems tend to be difficult to classify, since they are all capable of being significantly altered to suit a particular environment, the conveyor remains the most versatile, since its forms are too diverse to number.

Obviously, this list simply represents the major options available; it is not intended to be comprehensive. For example, one of the notable omissions is that of the ASRS. Although many people are aware of such systems as an increasingly popular means of automatically storing products or tools etc., these systems, when equipped with multiple load and unload points, may also be used effectively as a prime means of material transport. Indeed, such a combination of storage and material handling is appealingly efficient. Also, some of the assumptions used during the compilation of figure 6.1 may well not apply in certain circumstances. For example, conveyors may be costly to implement if they are required to cover long distances, with a capability of numerous routeing options; however, if the need is purely to transfer pallets of fixed dimensions etc. from one fixed location to another, a conveyor solution could be both reliable and cost-effective.

The selection of any one transfer medium is almost entirely dependent on the requirements exerted by the load to be carried. But once one selection has been made, further selections may well need to be compromised by the requirement of these later systems to integrate with the first. The following

Type	Nature of load	Load capacity	Speed	Flexibility of route	Cost	Overall versatility
AGVS	Discrete	High	Medium	High	V. high	High
Rail—guided	Discrete	High	High	Low	High	Low
Conveyor	Cont.	Low—md.	Low—high	Medium	Low—high	V. high
Robot	Discrete	Low—md.	Medium	Low	Md.—high	Medium
Gantry	Discrete	Low—md.	Medium	Low	Low—high	Low
Manual	Discrete	Low	Low	V. high	Low	V. high
Fork—lift	Discrete	Low—md.	Medium	High	Low	High

Figure 6.1   Categories of material handling systems

sections draw some general conclusions about the attributes and applicability of the main types of material transport systems.

6.3 Automated Guided Vehicle Systems

The AGVS is probably the most written about form of material transport to be added to the potential FMS design arsenal of high technology automation equipment. Although guided vehicles appear in a wide variety of different guises, the basic principle on which they operate is very similar. Essentially a guided vehicle is a battery-driven vehicle, intended to carry a quantity of goods from one location to another, automatically. They usually operate along a pre-defined route, or set of routes, which may permit a certain amount of flexibility in the actual journey made. Typically the vehicles are quite large, though some work has been carried out on developing smaller versions. A not unusual size would be approximately 36 inches wide, 54 inches in length and about 20 inches in height. The batteries on board the vehicles will allow several hours (usually between 6 and 9) of continuous operation between charging. Frequently, charging is fully automatic with the vehicle simply travelling to a special charge location, where contact between special pads on the floor and the vehicle allows recharging to occur.

Typically the vehicles ride on three wheels, two at the rear and one at the front. Drive may be provided by either the front or rear wheels, though the front is nearly always used to steer. Most vehicles these days are bidirectional — that is, they are able to travel both backwards and forwards. However, sometimes rearward motion is restricted to relatively short distances, perhaps for reversing into sidings etc.

While virtually all AGVs have microprocessor controls on board to make decisions regarding travelling, loading and unloading operations, there are a number of different guidance techniques. Without doubt the most popular appears to be wire guidance. A wire is buried in the shop-floor, usually about an inch below the surface, and inductive sensors, one on either side of the vehicle, allow it to track the wire and also to receive instructions from the central computer which is able to adjust the frequency of oscillation within the wire to send messages directly to a particular cart. It is possible to lay the wires in sections which can all be operated independently by the control computer. This allows traffic control of the whole system. However, free-ranging AGV systems are now being developed. Some of these use radio waves as the means of receiving positional and transport instruction data, while others use lasers to scan reflective sensors positioned around the factory walls.

The method by which loads are carried on vehicles varies quite considerably according to the application. Many vehicles are equipped with a hydraulic lift mechanism which allows them to carry, say, a pallet, on top, as shown in figure 6.2. The pallet location mechanism is able to move up and down as required, to collect or deliver a pallet. Some AGVs have forks, so that they may act in a manner similar to that of traditional fork-lift trucks. Others have the facility to load from the side, possibly both sides, and allow loads to be 'rolled' into position.

Although AGVS technology remains relatively expensive, it is nevertheless attractive to an increasing number of FMS designers, since such systems are extremely versatile, and are usually supplied with their own dedicated computer control system which only requires interfacing to the FMS computer for exchange of high-level commands. The real difficulty, if this mode of transport appears to be justified for an FMS, is that of facilitating the transfer of goods on to and off the AGV, especially if there is a wide variety of loads to be transported. The cost of docking equipment should not be overlooked. Incidently, neither should the tendency of AGVs to wear through almost any type of paint which is used to enhance the appearance of the FMS floor!

Certainly, the ease of extendability of the AGVS makes it particularly attractive. Even after a system has been installed, it is relatively easy to add either additional tracks or vehicles at a later date, as and when required. Also computer simulation is particularly well suited to optimising the design of automated guided vehicle systems.

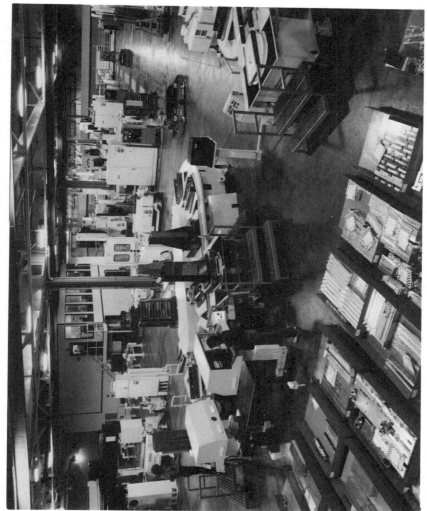

Figure 6.2   An example of an AGV (courtesy of Vought Aero Products Division)

6.4 Conveyors

Everybody is familiar with conveyors of one form or another. After all, they appear to be in use almost everywhere, in factories, airports, shopping centres etc. However, probably few people have taken time to think how many different types of conveyor there must be in existence. The main reason why conveyors are so numerous is that they remain a relatively inexpensive and reliable means of transporting loads. While complex systems might be controlled by a dedicated or part dedicated computer, most are run by programmable logic controllers (using ladder logic rather than high-level computer languages). Either of these may in turn interface to further programmable logic controllers and/or a higher-level computer for overall co-ordination. However, although the action of the conveyor itself is usually relatively straightforward, despite the fact that it could function in a wide variety of different ways, the overall control of the conveyor could be quite complex.

Unlike an AGV system where much of the intelligence is built either into the vehicle itself via the microprocessor control or the AGVS control computer, the intelligence for a conveyor system needs to be provided in another way. This is usually by the provision of a large number of sensors around the conveyor, all connected to the conveyor control system. These would be used perhaps to sense the number of a pallet, or maybe its precise location in order that it might be routed to the correct part of the system. It is this sensor and gating and/or routeing technology which frequently greatly complicates the implementation of a sophisticated conveyor system. Certainly such factors will have a considerable impact on the overall cost of the system, possibly being significantly greater than the cost of the conveyor hardware itself.

Certainly, conveyors exist in a wide variety of forms, for example, overhead monorail, carry and free, power and free, underfloor drag chain, floor slat, gravity feed, belt conveyors, plastic chain link, to mention only a few of the more common industrial systems. All of these systems are readily available in a wide variety of sizes, load-carrying capacities (in terms of both volume and weight) and also speed capabilities. However, despite the considerable variety in the physical characteristics of all these types of conveyor system, the high-level control characteristics remain remarkably similar. It is only when the actual control logic and sensor information is considered that the application-specific nature and complexity of the system become apparent.

A conveyor system cannot be as flexible as an AGVS, however where transportation tasks occur quite frequently, conveyors remain an attractive means of solving the work transfer problem. As a result of their mechanical simplicity, conveyors tend to be very reliable and with careful design, they can be made to transfer loads extremely efficiently and flexibly. As usual, one of the keys to success in any sophisticated system is the experience of the supplier together with the quality and applicability of the design itself, and where conveyors are concerned, experience is particularly relevant. Once a conveyor has been designed and installed it is extremely difficult, and certainly prohibitively expensive, to alter the fundamental operating characteristics.

Because of the wide choice of conveyor systems available, it is virtually impossible to give much that is meaningful in the way of selection guidelines for the potential FMS designer. All types of conveyors have particular advantages in certain applications but many of these overlap and in any one situation, any number of systems could be equally appropriate. The best route in these circumstances is to contact as many manufacturers as possible and see what level of correlation exists between their proposals. If a Systems House is being used to co-ordinate the supply of the work handling equipment, care must be taken to ensure that the relevant features desired of the system are clearly understood. Since conveyors have been used so widely for so long there is no shortage of basic design exerience, though the demands imposed by FMS do perhaps require a slightly different approach than is usually adopted.

6.5 Rail-guided transfer mechanisms

FMS designers over the past few years have developed what is effectively a compromise between a conveyor system and an AGVS. This is the rail-guided transport system, and is essentially a vehicle, very similar in appearance to that used in AGV systems, but instead of being guided by a wire buried in the floor, it is constrained to run on rails. Although the routeing flexibility of such systems is somewhat limited (they tend to be used only in straight 30—50 metre stretches), the advantages are quite substantial. First, such systems can be made to move quite rapidly, much more quickly than an AGV; and second, they are, in many respects, as reliable as conveyors, since on-board controls can be relatively unsophisticated and batteries are not necessary. All in all, if the FMS can be arranged in a straight line, a rail-guided transport mechanism could well be an attractive method of load transfer between bays.

The three techniques described above represent the ways in which the bulk of material transport tends to be achieved throughout FMS worldwide today. There are few flexible manufacturing systems in existence which do not use one of these techniques as the fundamental means of transferring material and equipment. This could be within one system or between several systems.

However, for smaller transportation tasks within subsections or subsystems of the FMS, another device (and some of its close derivatives) is frequently used: the device which has proven able to capture the imagination of authors, industrialists and film directors alike, namely robots — the industrial type.

6.6   Robots

Interestingly, one of the major outstanding problems with the application of robots is that despite the fact that they have been available in a variety of sophisticated forms for some years, little is generally understood about the difficulty which is likely to be found in their application, especially within industrial environments. Admittedly, this is a situation which is changing quite rapidly, essentially for two reasons. First, more engineering courses are including the teaching of relevant robotic knowledge, and second, many organisations are beginning to publicise the results of their earlier, and less happy experiences with robots, in order that the same mistakes need not be made again.

Robots essentially are CNC machine-tools specifically designed for work handling rather than material processing. They comprise various combinations of controlled joints, generally of two types: either pivots, allowing rotation, or ball screws (or similar), permitting linear movement. As control systems have become more and more sophisticated, a bias among manufacturers appears to be developing for the pivot joint, though while movement flexibility is enhanced by these, programming complexity is increased. Cartesian type robots, however, remain popular in assembly and disassembly operations, where linear motions are particularly important. Figure 6.3 shows some examples of both types of robot as supplied by Fanuc.

As with conveyors, there is a wide variety of different robots capable of transferring components in a flexible manner between what are usually relatively close locations. These could be just between one location and another, but are more usually between several locations within the robot's working area. Selecting the appropriate moves and actions is an integral part of the robot part program, and the relevant data which enables the robot to make some decisions is dependent on its environment.

In many cases, the key to the successful application of a robot is not the inherent design of the robot itself but the fact that the gripper has been designed to suit the application. Given this, it is hardly surprising that the cost of the gripper may utlimately represent a significant proportion of the robot system cost. Typically, the robot hardware itself (as despatched by the robot manufacturer) is likely to represent only 25–40 per cent of the total cost of the installed robot system.

Over the past few years, considerable advances have been made with computer numerical controls and this has greatly simplified the introduction of robots of ever-increasing sophistication, and ever-decreasing price. Indeed, this not only improves the programmability of the robots, but also the ease with which the robots and processing equipment may be interfaced and synchronised. However, although electrical interfaces between, say, machine tools and robots have improved, the problems associated with the mechanical interface remain as difficult as ever. The solution to handling a wide variety of differently shaped parts (components, tools etc.) with minimal equipment and changeover requirements is still extremely difficult, and often expensive to resolve.

Robots are now found in industrial concerns throughout the world. Japan leads in terms of the number of applications though there are some differences in definitions which make figures difficult to compare directly. When robots first appeared on the market, many people were under the impression that they represented the solution to most, if not all, difficult material handling problems faced by automation systems designers. This has certainly turned out not to be the case.

Robots have been shown to have both distinct advantages and disadvantages, depending on the application. Over the next few years, as experience grows and robots continue to become more sophisticated, the number and breadth of these application areas is likely to continue to grow. In the meantime, robots will probably remain particularly popular within paint-spraying and welding applications. Assembly and machine loading/unloading, where people originally thought that industrial robots would be most frequently applied, will probably remain in second place. Applications which are currently still under development, such as deburring, will probably soon increase in popularity.

Despite the advances that have been made with robots, they remain a relatively expensive means of handling loads. The reason why they are still popular is, of course, their inherent flexibility. As with a traditional CNC machine-tool, simply by loading a new program, the robot is capable of handling an entirely different task. This is obviously an attractive quality to an FMS designer, since flexibility in all the equipment components will help to provide flexibility within the system as a whole. As the software and the equipment it drives improves, so, to some degree, the mechanical problems become slightly more manageable. This is particularly the case if components require, say, reorientation between operations. A robot is an ideal device to handle such an operation, though at first sight this might appear to be a rather extravagant solution to a problem which could be solved with a simple turn-round fixture. This proves not to be the case since the robot is able to carry out many such functions, thus simplifying the control interface and the amount of component-specific tooling required. For example, in addition, the robot could perhaps remove the part from an input conveyor, load and unload it into all the processing equipment and eventually place it, in a pre-defined stack, on an output conveyor.

As with CNC machine-tools, the problem remains of how to generate good working robot part programs appropriate for the tasks in hand. This is usually both a complex and skilled task. Most robots have programs which have been taught manually with the use of a 'teach box'. This allows an operator to 'jog' the robot into a certain position, remember that position, and possibly the route to be used to travel to it, and the various functions to be carried out when it has reached its destination (open gripper, close gripper etc.). Much development is being directed to addressing the need to be able to program

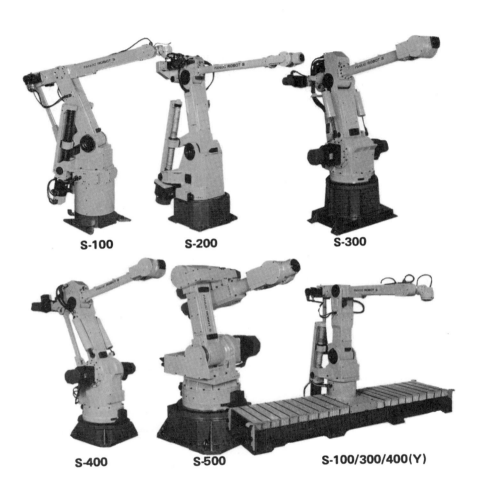

S-100   S-200   S-300

S-400   S-500   S-100/300/400(Y)

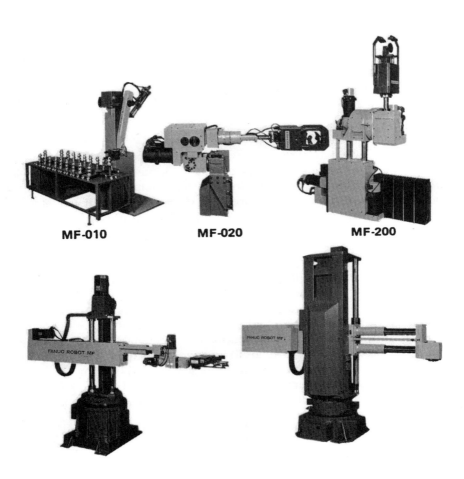

MF-010   MF-020   MF-200

Figure 6.3   Some examples of robots
(courtesy of Fanuc Ltd)

off-line, namely the ability to create working robot programs on a computer remote from the robot, in order that the programming task may to some degree be deskilled, and that the robot may be left working while the programming is taking place. The merit of this aspiration is discussed in chapter 12.

Considerable research is being directed towards the development of a standardised user-friendly robot programming language, which it is hoped all manufacturers will eventually use. However, for the next few years at least, it appears that robots will have to be programmed as they are now, by a skilled operator using a teach box.

6.7   Gantry loaders

Largely in an effort to obtain many of the advantages of robots and to combine these with the advantages usually associated with conveyors, the gantry loading system has been developed. There are many varieties of gantry loaders available, some of which are little more than a robot mounted on rails, but more often they are a robot arm which moves along a gantry-mounted rail, usually above the equipment being loaded. Such systems tend to be somewhat quicker than robots (which are often assumed erroneously to be as fast as their human counterparts), though slightly less flexible. Obviously, the routeing of the gantry loader is entirely fixed along the straight rail on which it is mounted. However, numerous stopping points can be established and, as such, gantry loaders can be used, quite successfully, to simultaneously tend a number of different processing stations (see figure 6.4).

Such systems are usually controlled by a programmable logic controller. While gantry loader hardware tends to be quite inexpensive when compared with that of robots, the problem of gripper design is much the same. In certain applications where the full motion flexibility offered by robots is not required, gantry loaders can represent an efficient and cost-effective solution.

6.8   Manual material handling

The most inexpensive form of material transport (in terms of capital investment rather than perhaps running cost), and probably the most versatile, remains that of the manual operator. This is especially the case when the operator is assisted by carefully designed automation equipment, for example, at one extreme, equipment such as a fork-lift, while at the other, it could be special equipment to assist with complex component datum or inspection operations. In terms of loading and unloading awkward components into complex jigs and fixtures, there is very little to compare with the capabilities of a skilled operator. Assembly operations, in particular, tend to

Figure 6.4   An example of a gantry loader
(courtesy of General Electric (USA))

be very intricate and hence are not even easy to automate, never mind be
more efficient than a manual system on an operation-by-operation basis.
Where, of course, automated equipment is distinctly advantageous is where
continuous (24 hours a day), accurate and consistent operation is required.

Although there is a great deal of discussion about totally unmanned
factories, and indubitably FMS is a step towards this concept, the development of viable demanned factories is far more likely. To develop the software
needed to eliminate operators completely is an awesome task. Health and
safety regulations are unlikely to permit complex automated facilities to be
run by just one operator, usually two will have to be present at all times.

But if two operators are present, it is then important that they are utilised
by the system in an optimal manner — that is, by making important or
perhaps complex decisions which might otherwise be difficult to automate.
This means identifying those operations which are key to the success of the

overall system and making an operator responsible for overseeing the correct completion of that task. However, operator attendance at a particular time should not be designed into the FMS as a mandatory requirement, since in a demanned environment such attendance cannot be guaranteed, and the production process should not be delayed while waiting for an operator. These are unlikely to be purely loading and unloading, though these might well be included in the list in certain instances. More likely such tasks are long-range scheduling decisions, recovery of the system or an element of the system after an error has occurred, and probably component inspection. Therefore if, for example, the inspection task can be combined with more mundane tasks without impacting adversely on the overall efficiency of the system, the considerable development costs for control software, jigs, fixtures and transportation could well be avoided.

6.9 Buffer stores

The usually crucial need to minimise the work in progress within the system impacts considerably on the design of the material handling system. Buffer stores should only be used to negate the effect of poor equipment utilisation and/or imbalance due to a particularly awkward work schedule, or maintenance, or unavailability of transport etc. As such, buffer store sizes should be strictly limited, and the control algorithms of the FMS should avoid the use of the stores whenever possible.

Where conveyor systems are used, it is popular to have accumulation lanes for use as on-line buffers. Where AGV systems or rail-guided systems are being employed, static buffer stores appear to be the most favoured technique for providing work-in-progress buffer stores. Unfortunately these can frequently be costly if special fixtures are required, owing to the fact that the fixtures have to be duplicated to meet the requirements of, for example, the load/unload stations. Indeed, this is a further consideration to be taken into account with fixturing. Since fixturing is almost invariably expensive, the last situation one would want to generate is one where all these expensive fixtures are gathered in the middle of the FMS, together with expensive work in progress. Ideally, there should only be as many sets of fixtures as the number of parts one would typically expect to be in the FMS at any one time, perhaps with some allowance for worst-case situations. But then there is the risk that fixtures will need to be changed from part to part, as the batch progresses, a situation which can rapidly become chaotic if work-in-progress stores are too large, or their contents too loosely controlled — not to mention the adverse impact such a situation might create at the load/unload stations.

A tacit advantage with most fork-lift, conveyor and AGV systems is that, unlike some automated systems, the work in progress is usually distributed around the shop-floor, of a defined maximum capacity, and clearly visible for all to see. This includes upper management, who will probably be among the first to notice when the buffer stores seem to be too full.

An ASRS is essentially a computer-controlled, completely automatic warehouse. Parts are delivered, possibly by any of the above-mentioned material handling systems, and storage is entirely under the auspices of the ASRS control system. These systems tend to be quite expensive (though they are available in a wide variety of different sizes and levels of sophistication), and have the considerable advantage of being self-contained, and capable of handling a large number of parts within a fairly small area of floor space. This is largely because they have relatively narrow aisles between the storage racks, which are then densely packed from floor to ceiling, which, incidentally, could be quite high in especially designed high bay areas (see figure 6.5).

(a)

Figure 6.5   (continued overleaf)

(b)

Figure 6.5   Examples of ASRS: (a) a high-bay ASRS; (b) a mini-load
ASRS for storing tools (courtesy of Eaton–Kenway Inc.)

One of the main disadvantages with these systems is that if they do break down (which admittedly is fairly rare, since at least the better-made systems tend to be extremely reliable, for example, a highly reputable French manufacturer quotes up-times of 98 per cent), it can be quite difficult to operate the system manually, not only because the systems have not been designed with manual operation in mind, and hence can be somewhat daunting to the operator, but also because it may well be extremely difficult to establish where the required parts are being kept. This is because the ASRS computer allocates part positions usually on the basis of a fairly sophisticated algorithm to ensure the most frequently needed parts are closest to the system exit, and since it could well be that this data has been lost, the storage locations might not be readily obtainable. This would necessitate a painstaking search of the entire contents of the store to establish the precise location of part, tool, fixture etc.

Also, apart from the high cost of these systems, the contents of the ASRS are not easy to inspect. For example, it is difficult to ascertain whether the system is completely full, half full or nearly empty. Certainly this information is readily available to those who know how to obtain it. But it is much easier for such a facility to be mis-used, and for a problem to go almost unnoticed. For this reason therefore, it is desirable to have work-in-progress stores clearly visible. The fullness of these, together with the level of activity of the process equipment, gives a fast and fairly accurate impression of the utilisation and efficiency of the FMS.

6.10   Selecting the correct material handling systems

The preceding discussion has outlined the major classes of material handling systems available to FMS designers. However, it is difficult, if not impossible, to provide much more than broad guidelines for the selection of a material handling system for a particular FMS. There are a number of considerations which have not been clearly identified which are likely to be of value. An effort has been made to create below a generalised summary of the most important of these. It should be appreciated, however, that in many instances the influence of various other factors (such as already installed systems and cost) might significantly bias the selection process.

If the distances involved within the FMS are relatively large, and the components also relatively large, it is likely that the event arrival times will be relatively large. In such an instance an AGV system might well prove appropriate, especially if there are many load/unload locations, and the connection routeings are near random state.

If event arrival times are short (that is, processing cycle times are short), but otherwise the circumstances are essentially as described above, then some sub-batching may be required. This could be achieved by loading a number of parts on to one pallet (thus reducing the number of transportation operations which need to be carried out to move a certain number of parts). Alternatively, two pallets could be transported at the same time.

If the component transfer rate is high, then an intelligent (that is controlled) conveyor system is more likely to be appropriate, especially if the routeings are not too random and the distances involved are medium to large. If pallets are used on the conveyor, then an element of sub-batching may again be desirable to ensure an effectively continuous flow of work to and from the subsystems.

If processing times and transportation distances between load and unload locations are relatively short, then either a robot or a simple conveyor system would probably suit the application. If the load and unload locations are fairly numerous, and especially if a certain degree of intelligence is needed to determine the destination, then an intelligent robot is more likely to meet these requirements than a conveyor system.

If the parts are relatively small, the process workstations can be located in a line and if transfer times need to be fairly small, then a gantry system might be desirable. However, if in the above case the parts were quite large, then a rail-guided transfer system would be preferable.

The ultimate goal of all material handling systems is to ensure that sufficient material arrives at the workstations just as it is required, with a minimum amount of work in progress. However, for a variety of reasons (such as cost constraints) one will almost inevitably have to settle for something less than the ideal, but that does not mean that it is necessary or desirable to lose sight of the ideal.

It is desirable that the number of different material handling systems within any one FMS is minimised. This greatly simplifies installation and commissioning, and impacts favourably on both design and maintenance. It is unfortunately often the case that more than one type of material handling system will be needed. For example, in an FMS equipped with a rail-guided transport system, it is likely that another device will be needed to transfer the tools to the machine-tool tool-carousels. This was the case in General Electric's Erie FMS, where a robot was used to carry out this task. Conceivably the tools might even need to be loaded manually, in which case the FMS designer has to accommodate this requirement in a safe and efficient way. In a conveyor-based FMS, it is likely that components, tools and other equipment will only be delivered relatively close to the process workstations. A robot, a gantry or an operator is likely to be needed to complete the transfer into the equipment.

In machining centre-based systems, the incorporation of pallet changers, automatic tool changers and tool-carousels has tended to simplify the issue of mixes of transport systems, simply because these systems arrive as an integral part of the machine-tool itself. However, when considering these various options available with processing equipment, one should bear in mind that these systems are being supplied as a machine-tool manufacturer's solution to a perceived problem. This may or may not mean that the equipment is going to solve the particular material handling problem being considered. Indeed, even if it does, it could be that the equipment complicates the issue by perhaps being difficult to interface.

If, as is likely, it is unavoidable that there are different types of transport system within the same FMS, then it is important that the number of transfers between different systems is minimised. Interfaces, whether mechanical or electrical, are always difficult and expensive to engineer. For example, if one is transferring components of a relatively small cycle time, one could transfer these individually and ideally very quickly from one workstation to the next. Alternatively, one could sub-batch them, say on to pallets, and then transfer the pallets relatively slowly between workstations; however, this would result in slightly more work in progress in the FMS and the overlap between different operations would be slightly less optimal. It might also significantly reduce the amount of expensive equipment such as AGVs, required. Certainly, decisions of this type can only be made on the basis of a careful analysis of the components and their value, and of the FMS components themselves; thus computer simulation (see chapter 7) is an essential part of such an analysis.

If the FMS being considered is one of a number which are being planned, it may well be necessary to select from the outset the transportation system which will be used to support the activity of all the flexible manufacturing systems. This will complicate the decision-making process, not only from a technical viewpoint but also from a commercial one. Issues such as the type of parts being manufactured in all the systems, and all the various tooling, fixturing and storage requirements, will need to be considered at the same time. Also, whether there is to be one cell which expands into a larger system or several systems which are eventually to be linked together, will need to be considered. In the former instance, a conveyor might be appropriate if the components have the correct characteristics. In the latter case, an AGVS would almost certainly be preferable. If the distances to be travelled are quite large, a radio-controlled AGV might be more effective than a wire-guided system.

6.11 Concluding comments

The number of material handling systems and the viable ways in which they may be combined are numerous. The only way to ensure that as many options as possible are considered is to monitor all the available technical journals and consult all possible suppliers and manufacturers of appropriate products.

Before committing oneself to a particular system or combination of systems, much thought must be given to the layout of the FMS. This not only means establishing the physical distribution of the processing equipment itself, but also the location of such features as power distribution lines, pneumatic supply lines and possibly even the swarf removal system since this could perhaps be via an AGVS. The location of machine-tools and co-ordinate measuring machine foundations away from busy transport points could also be a relevant factor. If the FMS is to be installed in a 'green-field site' with a purpose constructed building, then the constraints within which the system must be sited might not be significant. However, if the system is to be located within an existing facility, such factors as structural pillars might severely restrict the layout options.

Also, the impact of the requirements of maintainability, operator safety, operator access, extendability, and disruption to an existing, and fully operational, production facility, should not be overlooked. If the part mix is likely to alter substantially during the life of the FMS, it may well be worth considering the ease with which equipment could be relocated to suit the new components' requirements. The influence of many of these factors is discussed in chapter 11.

After all the relevant issues have been investigated, it is likely that a short list of viable material handling system combinations and suppliers will evolve. In these circumstances, if, for example, cost considerations do not dictate a definite answer to the problem, such factors as maintenance expertise, vendor proximity and experience, technical risk etc. should be taken into account to help make the decision. Possibly the feasible layout/material handling options will be based on entirely different philosophies. For example, an AGV-based system might seek to minimise the inter-workstation material transfer, while a conveyor system might wish to maximise such transfers. If all other factors appear effectively equal, it might be worth considering the influences of some other factors such as:

- If one part of the system fails, how easy is it to maintain operation of the remainder of the system?
- Is manual intervention straightforward for, say, planned and unplanned maintenance?
- How reliable is the system likely to be? For example, are any innovative, unproven sensor technologies incorporated in major functional areas?
- How complex is the control system required to be?
- What are the guarding/safety implications?
- If an AGV system is being considered, what is the effect of recharging times, and are recharging stations able to be located close to busy traffic areas?
- What is the effect of traffic build-ups within the system — that is, how does the transport system (and indeed the rest of the system) perform in the worst-case situation?
- What is going to be the impact of the material handling system selection on the tooling, jig and fixturing requirements of the system (after all, these could be a very significant cost item, and choosing a particular material handling system might significantly reduce the amount of specialised tooling etc. required)?
- How much floor space will the system occupy?

Factors such as these, and the list is by no means complete, might well be sufficiently relevant to clarify an otherwise complex material handling system decision.

With regard to the order in which workstations should be placed within a layout, there are only a few basic rules worthy of note which could have a significant impact on the design and type of the material handling system.

Obviously, the magnitude of heavy traffic areas should be reduced to a minimum, so any workstations between which there are a particularly high number of transfers should be placed close to each other. Also it is often helpful to group similar workstations together, and it is desirable to have work flowing in a consistent direction throughout the system. For example, if there is a line of machine-tools, one end should, whenever possible, be the system entry, first operation point, the other end the finish point. Then as work passes down the line, it will always be nearing completion. This could be of particular value in an assembly system where more parts enter than leave, since it will reduce the number of transfers later in the system, and if a conveyor is being used, this could substantially reduce the investment required for the material handling system.

When selecting a material handling system, one important feature which should be taken into account is that of error handling. Everyone would like to believe that the system they have designed is unlikely to go wrong, but it is unfortunately highly likely that one day something will go very wrong.

For example, a robot might drop something or a pallet might end up at the wrong destination. While to a large degree the ability of the material handling system to recover from such disasters is going to be dependent on the quality of the FMS control software, an element of the difficulty of recovery is dependent on the inherent design of the material handling system. For example, if one is using an accumulating conveyor system with pallets, and it is discovered that the middle pallet in the queue is incorrect, it might be extremely difficult to extract it, both physically and in control terms. Whereas if the same situation arose within an AGV system, the queue might be a static buffer, and hence the removal of one pallet could be quite straightforward.

Finally, when making a decision on material handling systems within an FMS environment, it is essential, as with all other aspects of the system, that all the relevant advantages and disadvantages are taken into account at the outset. A decision made hastily on incomplete data could eventually represent an extremely expensive mistake.

Some of the considerations appear almost trivial, however they are still links in the chain, and all links have to be present if the chain is to be of any use. For example, if robots are being used, gripper design could be both expensive and difficult, even with some of the CAD facilities now available. Standardisation of, for example, tool holders can assist with minimising the effect of this problem. The magnitude of the robot programming task should not be underestimated in terms of the quantity and quality of resources required.

Problems of component or tooling orientation should not be ignored. Robots certainly are extremely flexible in this respect, but often under-utilised, because they are serving other workstations. If a robot is being used at a workstation to load components, it could perhaps also be used for other tasks such as loading tools or rotating components. Such applications might not be optimal, but if the equipment is already in place, the incremental cost of using it for an additional task is small. Such issues could well affect the equipment selection process. For example, a gantry loader might necessitate a dedicated turnover fixture whereas a robot might be able to handle the whole operation with a minimum of additional fixturing.

However, robots are usually not as fast as many people believe. Care should be taken to ensure that unnecessary idle time is not imposed on costly capital equipment, such as machine-tools, purely because a robot is having to use flexibility to make up for inadequate material supply. Bin picking is a good example of this phenomenon. Much research is currently being devoted to automating this particularly difficult task. However, ideally, workpieces should not be allowed to lose their orientation once it has been established. Admittedly, there are some instances where this is virtually inevitable, for example, when small stamped components are ejected from a press at such a high speed that it is difficult to stack them. But if bin picking, or simply obtaining the next part for processing, is required, it is desirable to ensure that as many of these tasks are carried out in parallel with the processing operations.

If all these factors are taken into account, then the probability of a successful FMS being implemented is increased substantially.

7 Computer Simulation

7.1 Introduction

To many people, the computer simulation of manufacturing systems remains a mystery. However, if simulation had been available during the early days of FMS, it is likely that many of those systems would have been more successful. Recent advances in computer simulation, many of which will be described within this chapter, have made this technology both effective and readily available to manufacturing system designers.

This chapter is split into three sections. The first is intended to introduce the possibilities offered by computer simulation; the second discusses microcomputer-based simulation packages; the third section discusses when and how simulation could be used during the implementation of a typical FMS.

7.2 Simulation for FMS design

The computer simulation of sophisticated automation systems in both the discrete parts and process environments has been relatively common since the late 1960s. Usually such exercises were carried out to assist with the identification of system bottlenecks, and the utilisation of fixed resources (machine-tools etc.) and variable resources (that is, operators, tools, transporters etc.), both of which were usually of a finite capacity. This type of analysis would then give the designer a reasonable degree of confidence as to how well the system would work, together with valuable information about factors such as work in progress, production rates and the impact of equipment failures etc. This type of modelling exercise was proved, by the few people who applied it at that time, to be particularly relevant to the FMS environment.

Typically, this type of performance modelling exercise was carried out on a mainframe computer using a fairly high-level language. While these languages were indubitably highly sophisticated, they did tend to have disadvantages, which in many respects is reflected by the fact that they were the forerunners of the systems available today. For example, the model itself usually had to be programmed using the particular syntax relevant to that language. At best, this turned out to be a laborious and time-consuming process, one which more often than not had to be carried out by an extremely skilled individual, of which there was usually a shortage. Frequently, the models which resulted could only be understood by the people who created them (that is, the programmer) and one was never too sure that the programmer had created a model which was really simulating the designer's proposed system. Certainly the customer, who was actually sponsoring the design effort, was often the last person to be sure that the model's logic reflected the specified FMS requirements. As for the substantial amounts of printed results which, more often than not, were generated — these, if anything, simply tended to aggravate the situation.

However, during the past few years, notably since the late 1970s, the original high-level languages have been improved considerably in terms of the functionality which they offer. The first stage of this evolution was the more traditional, mainframe simulation systems being interfaced to a variety of graphics and animation packages, and this was rapidly followed by the development of graphics pre-processors and post-processors. These greatly improved the ease with which models could be created and their output interpreted. Many have had constructs or special subroutines added to facilitate the modelling of, for example, material handling systems. But there remained a significant barrier to the frequent application of even these systems, and this was essentially because of their lack of 'user-friendliness'.

Recently though, the advent of sophisticated colour graphics together with the numerous and substantial advances which have been achieved with computer hardware has allowed software developers to create even more sophisticated packages which are considerably easier to use, and which also contain such features as animation. No doubt anyone who has had the opportunity to use examples of such a package would agree that 2 minutes of animation, followed by the display of relevant graphs, is far more meaningful than many thousands of pages of computer print-outs.

However, even with the relatively sophisticated output analysis packages generally available, the major fundamental problems remained, namely that an 'expert' was needed to create and interpret the model, and also that a substantial amount of time on a large mainframe computer was needed actually to run the model.

Indeed, it really does seem rather incongruous, in an age of rapid information transfer and significant increases in the computing power available to a user, for example, in a relatively inexpensive microcomputer, that designers of manufacturing systems have not, until very recently, been able to create and interpret models of manufacturing systems on their own. Of course, to develop such a modelling system is not a trivial task. To be able to create and run a model of a complex automation system does require a highly sophisticated modelling system, capable of enabling the user to define the planned facility, to input design parameters in a convenient way, to run the model and to analyse the results. Additionally, facilities would need to be provided to allow the user to change operational parameters, and then re-run the model. Furthermore, to be of real value, all this would have to be provided in a form which does not require many months to learn. Suffice to say, such systems do indeed exist, and they need not be particularly expensive to purchase.

So, while in the past the growth of the flexible manufacturing and advanced automation markets substantially benefited the proponents of a variety of complex operations research techniques, the overall result was in fact to stimulate the growth of substantially more 'user-friendly' simulation systems. As such, bureau service simulation, while still having a useful role to play in FMS design, appears now to be a declining business.

But, even all these technical developments do not represent an entire solution to the problems confronting factory simulation. For example, there is also the issue of the level of credibility which can reasonably be given to the results. Highly skilled and respected production engineers will often point out that complex manufacturing systems were built long before computer simulation was developed. A significant proportion of these had worked (or at least one did not hear too much about those that did not!), so why should it be necessary to simulate now? Of course, this is a somewhat jaundiced view, but alas it is one that is shared by most of us to a greater or lesser degree, about some aspect of our lives. Once one becomes even slightly set in one's ways, the prospect of having to change is not particularly appealing.

Probably the only answer to the issue of the credibility of simulation is that of time. As more people become familiar with the substantial benefits which computer simulation can offer, and there are more well-publicised examples of where simulation was used to advantage, so simulation will become a more acceptable technique. The influx of younger engineers, more familiar with the use of computer systems, will inevitably help this transition.

For the time being, however, it often continues to be difficult to convince hardened production engineers of the benefits which can be gained from computer simulation as opposed to 'hard modelling' (that is, the use of a scale layout, pieces of paper and drawing pins to represent pallets, people etc.). Fortunately, however, time, desk-space and drawing pins are not usually available in sufficient quantities to permit anything other than a relatively trivial subset of a complex automation system to be modelled. Anyone who has lived through such a 'hard modelling' exercise will no longer doubt the value of computer simulation. Similarly, anyone who has had to resort to 'remote' mainframe simulations of a system design for which they are responsible will not doubt the value of sophisticated user-friendly microcomputer simulation packages.

A further issue which to some degree continues to haunt computer simulation is that of funding the analysis. Simulation, especially when carried out on a remote mainframe computer, is likely to be relatively expensive. Justifying simulation expenditures perhaps when one is only at the essentially formative stages of a project, is unlikely to be easy. Once again, one frequently comes up against the opinion that if simulation was not necessary when one replaced one machine-tool with another, why should the incremental expenditure be necessary now? While there is little if any justification for such an opinion, the fact remains, it is likely to arise on some occasions during the implementation of an FMS, and will represent one of the many day-to-day problems which need to be addressed, tactfully, by the Project Manager.

An effective way to overcome this issue is to ensure that all the interested parties meet with some people who have used simulation, ideally as a design tool to assist with the implementation of an FMS. If the same people previously also have experience of implementing such systems without simulation, this will obviously be an added advantage. Without doubt, in circumstances such as these, the probability of the virtues of simulation, especially on a microcomputer, not being wholeheartedly accepted are extremely small.

However, the strategic problems faced when carrying out computer simulation exercises remotely, might still have to be faced; for example, the fact that it is difficult to exert much control over the modelling process itself. While it is unlikely that a 'reputable simulator' would produce an inaccurate model intentionally, there is, as mentioned above, little guarantee that the model is a true representation of the planned facility. The modellers themselves are unlikely to be experts in the type of facility being designed. So the key issue is the amount of knowledge which can be exchanged between the system designers and the simulators. Inevitably there will be some loss of data within this communication process. No matter how sophisticated the model is, it will never be a perfect representation of either the real system or the intentions of the FMS designers. To whom the blame for this should be allocated is largely immaterial, the fact is that the discrepancy will exist. Similarly, once the model has been run, the results have to be interpreted by the simulator and passed on to the FMS designers. Again, there is bound to be a slight loss of data within these communications. The project engineers should be aware of these problems, take every precaution and ensure that the overall effect is minimised.

The final major drawback with remote simulation exercises is the difficulty of accommodating changes. Certainly, design changes within an FMS are inevitable. Indeed it is highly desirable that a number of viable system alternatives are considered, and the most appropriate combination selected. For example, if one decides to juxtapose two workstations, ideally one would like to be able to simulate this and see the effect immediately, or at least as near immediately as is practicable. However, if the model is remote, this process could well take, say, a week; and then there is the risk that the model was constructed in such a way that it is not straightforward to make such a change. This would then prove expensive as well as time-consuming, and could be quite demotivating for both FMS designer and modeller.

Fortunately, as intimated earlier, there is now a way of avoiding many, if not all, of these problems, at least for the majority of most FMS simulation analyses, and that is by using one of the many microcomputer-based simulation packages.

7.3 Microcomputer-based simulation systems

The development of microcomputer-based simulation systems has largely been made possible by the rapid advancement in the power of the computer hardware on which they run. For example, it is now possible to run a sophisticated FMS simulation system on a battery-powered portable micro-computer. Admittedly, such a model might well run rather slowly, but as is the case with the computers controlling FMS, it is not the speed of the process itself which dictates the overall efficiency of the system. But, there is the possibility that a model created on a microcomputer-based system might not be as comprehensive as one which is created using a sophisticated mainframe simulation package. However, the likelihood of this is continually being reduced as more and more sophisticated simulation packages become available, and also as more powerful microcomputer hardware becomes available.

In fact, in the not too distant future, it is likely that a microcomputer-based simulation system will be just another of the tools used by any technology conscious engineer, as is already the case with spreadsheets and word-processors. However as these systems become easier to use, a new risk is created.

For example, it is quite likely now that an enthusiastic FMS Project Manager, rather than consult an outside supplier for a simulation, might instead decide to purchase a microcomputer-based simulation package and have the modelling carried out by a member of the Project Team. This is certainly highly commendable. The probability of a slightly less comprehensive model being created will be far outweighed by the level of knowledge available to the model's creator and the substantial increase in 'turn-round' speed of different model configurations. Nevertheless, the possibility still exists of an inaccurate model being created, and/or the results being interpreted incorrectly. The risk is enhanced by the fact that this could be the modeller's first opportunity to use the package, and that the model itself might not be examined as closely as would be the case if the work was being carried out remotely.

These are issues which must be addressed by the Project Manager: first, to ensure the modeller is adequately trained; second, to ensure the results are carefully scrutinised and interpreted; but above all, to ensure that the expectations of both the model and the modeller are not unreasonable. If these precautions are not taken, the simulation exercise could yet do more harm than good.

There are a number of microcomputer-based simulation packages on the market, and indeed many of the so-called mainframe simulation packages are now available in a form that will run on a microcomputer. A small selection of the many packages now available is given below:

FIST	FMSSIM	ECSL
GPSS HEI	HOCUS	MAP/1
MAST	MODELMASTER	PCMODEL
SIMAN	SLAM II	SEE-WHY
SPEED	TESS	

Certainly, this list is far from complete, and is not intended to suggest that these are necessarily the best systems currently available. It does at least give an indication of how many different systems there are available. Obviously, each has its merits in certain application environments. Some are rather easier to learn to use than others, and some are rather more comprehensive than others etc. Obviously one has to make a compromise between the relative importance of all these factors, prior to making a purchase, or a leasing decision.

Many of the systems mentioned above produce excellent graphics, and there is no denying that a picture is worth a thousand words. In fact, this phrase when applied to simulation probably represents a substantial understatement, essentially for two reasons. Not only is it much easier for the modeller to build and interpret the results of the model if some sophisticated graphics are available, but also the value of some realistic graphics when explaining how the proposed system is expected to work, to, say, members of upper management, perhaps comprising the Steering Committee, or even a potential customer, is quite incalculable.

However, while there are many systems which are able to produce graphic outputs (indeed several of the above list), few also provide the facility for graphics input, namely the ability for the modeller to generate a simulation without any programming (that is, to permit the modeller to work in an entirely programming-free environment).

Such a system is commercially available. It was created with a view to overcoming the difficulties mentioned above, and is called *Modelmaster*. It has been developed by the General Electric (USA) Corporate Research and Development Centre in Schenectady, New York. Although it is not intended to expound upon the virtues of any one particular simulation system, it is appropriate that an insight should be given as to how easy it now is to carry out a simulation analysis of a proposed system, such as an FMS. Therefore the following section of this chapter is devoted to a description of how Modelmaster, a package which is representative of the type of sophisticated microcomputer simulation system soon to be generally available, would be used in such an exercise.

7.4 The 'Modelmaster' simulation package

The history of Modelmaster's development in some ways reflects the confusion which is often associated with computer simulation throughout industry worldwide. It was found that within the many General Electric manufacturing plants, computer simulation was becoming more and more popular. This was reflected in, among other things, the ever-increasing variety of simulation systems which could be found to be in use. So with a bold brief, a relatively small team of experienced, highly skilled people set out to develop a new simulation system. The team had four distinct goals in mind for the system:

(1) It had to be comprehensive, because General Electric's manufacturing environments are very diverse.
(2) It had to be simple to use.
(3) It had to provide the user with a completely programming-free environment.
(4) It had to make maximal use of colour graphics for both input and output.

As such, the Modelmaster simulation package is designed to enable people who have little or no previous experience of computer simulation to create and simulate models of complex manufacturing systems. At the same time, it ensures that these activities may be carried out both quickly and efficiently, while minimising the risk that an inaccurate model is being created.

The package, which was originally intended to be used only within General Electric, essentially comprises four highly flexible simulation modules: one for serial processing, one for assembly loops, one for the job shop and one for material handling systems. These modules were unique in as much as they all included an extremely user-friendly, programming-free environment, comprising both graphic and menu-driven input. This enabled the user to create simulation models which could be run in a manner which was totally transparent to the user, and provided a number of different levels of graphical output.

From these four basic modules, hybrid systems of more general application were evolved. For example, by combining the serial processing and assembly loop modules a system particularly well suited to the simulation of assembly facilities was created. Similarly, by combining the job shop and material handling modules, a system particularly appropriate to the modelling of flexible manufacturing systems resulted. After these packages had been used within General Electric for some eighteen months, it was realised that a commercial version could be developed quite easily. This work was duly carried out and resulted in the launch of the package as a microcomputer-based simulation system during late 1985. The recently upgraded package contains the following functions:

(1) *Menu-driven graphical layout*

- System entry and exit
- Workstation location
- Buffer store location
- Job types and sequences
- Multiple operations per job
- Transporter paths and aisles

(2) *Forms-type data entry*

- Batching and splitting of jobs
- Assembly of job types
- Probabilistic batching and scrap
- Resource/shift management
- Planned and unplanned downtime
- Scheduled raw material arrivals
- Initial steady-state loading capability
- Multiple transport systems
- Automatic calculation of transport times
- Auto queue ranking (FIFO, LIFO etc.)
- Queue capacity constraints
- Set-up operations
- Full retrospective edit capability

(3) *Simulation program and reports*

- Specification of run duration
- Batching for statistical analysis
- Auto-removal of spurious start-up data
- Traces for specific time periods
- Average, min. and max. queue contents
- Average, min. and max. workstation use
- Resource and transport utilisations
- Multiple run batching

(4) *Graphical output*

- Queue content bar charts
- Resource utilisation pie charts
- Workstation utilisation Gantt chart
- System animation

It is clear from the above list that Modelmaster is a comprehensive package. The following, relatively simple example, demonstrates how easy it is to use.

Consider the simulation of a relatively simple flexible manufacturing system. The first step is to prepare data for input into the modelling system. This will comprise both equipment layouts and component-related data. The equipment layouts would need to be accurate, scaled representations of the layouts to be considered. The component-related data includes work schedules, process routes and operation data. This could all be extracted from the data sheets which, hopefully, would have been prepared as described in chapters 4 and 5. It is imperative that this data collection phase is carried out methodically and comprehensively. As with all simulation systems, the 'GIGO' rule applies (Garbage In, Garbage Out!).

The layouts are then used as the initial input to the modelling system, a scaled interpretation of the proposed layout being drawn on the microcomputer screen either with a 'mouse' or simply the standard keyboard. 'Rubber banding' techniques are incorporated into the graphics routines to facilitate this process. The end result is a display of the facility layout, with distances scaled to represent those which would exist in the actual installation (see figure 7.1).

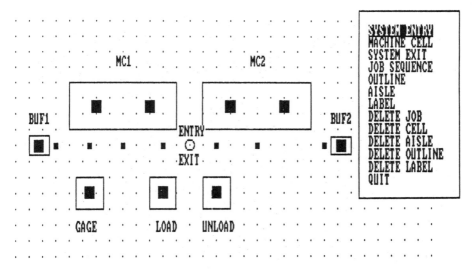

Figure 7.1 Simulated FMS functional layout

In this case the flexible manufacturing system comprises four process workstations (say, machining centres), these being of two different types, separated into two bays called MC1 and MC2. There are three other bays within the FMS: one associated with the gauging of components (that is, GAGE), and two responsible respectively for the loading (LD) of parts into, and the unloading (UL) of parts out of, the flexible manufacturing system. The workstation in these bays is also required to handle the mid-process refixturing of parts. Also within the system are two on-line buffer storage bays (BUF1 and BUF2) which serve the whole system.

All these bays are then linked by a material handling system, typically either a rail-guided transfer mechanism or an AGVS. The route of the transport system is input to the model by pointing the 'mouse' at the appropriate workstations, in an order which reflects the processing requirements of the parts, and 'clicking' the button on the 'mouse'. If required, the 'rubber banding' capabilities of the graphics input package could be used to trace out the precise path to be followed by the transporter between the various input and output queues.

The remainder of the information required to run the simulation model is then input with the aid of a sophisticated forms-entry system. Most of the questions on the menus are formulated by the package on the basis of the system layout previously input. At all times, a context-sensitive help facility is available, so if the user runs into difficulty, a text help file can be 'pulled down' on to the screen.

In addition to inputting all the required component, schedule and routeing information, constraints on the maximal number of parts which could be accommodated at the various queue locations, were input as follows:

$$
\begin{array}{ll}
\text{MC1} - 2 & \text{MC2} \ - 2 \\
\text{BUF1} - 5 & \text{BUF2} - 5 \\
\text{GAGE} - 1 & \text{LD} \quad - \text{ unlimited} \\
 & \text{UL} \quad - \text{ unlimited}
\end{array}
$$

The system was to process three types of job, with a total requirement for 80 of each part type to be manufactured during each week. The raw material to be processed into the finished parts was assumed to arrive at the FMS at regular intervals throughout all the three shifts. The processing sequences for the three jobs was as follows:

JOB1 — LD, MC1 or MC2, GAGE, MC1 or MC2, GAGE, UL
JOB2 — LD, MC1 or MC2, LD, MC1 or MC2, GAGE, UL
JOB3 — LD, MC1 or MC2, GAGE, MC1 or MC2, LD, GAGE, UL

Finally, a number of assumptions were made about the operation of the system:

- Loading, unloading and refixturing would each take 10 minutes.
- Workstation cycle times would vary between 20 and 40 minutes, depending on the operation.
- Part gauging would take 10 minutes.
- The overall length of the system was 100 feet.
- The single transporter, with the ability to carry one part at once, would travel at a speed of 100 feet per minute.
- Each simulated time unit would represent 1 minute.

The appropriate queues were then pre-loaded to half their capacity at the beginning of the simulation, thus ensuring that a steady-state situation will be obtained as soon as possible. Then the simulation option is selected on the main menu, and the model is run, in a manner totally transparent to the user. No generation of computer code has been necessary, and a lengthy learning period is not required — features which result in a considerable increase in efficiency.

In this case, the flexible manufacturing system was simulated for the equivalent of three eight-hour shifts, for five days a week (not really enough for a real simulation analysis, but adequate for the purposes of this text). However, the load and unload bays only had an operator available during the first two shifts of each day. In addition, it was assumed that all the machining workstations would require an average of one hour's preventative maintenance each week. This work would be initiated on a random basis.

Once the simulation model has been run, the results may be obtained in a variety of ways; for example, in the form of tabulated reports, or alternatively as graphical output, examples of which are shown in figures 7.2 to 7.4.

Figure 7.2 shows four graphs of queue contents against time for the system entry (LD), one of the buffer storages (BUF1), the input queue for the first machine bay (MC1) and, finally, the input queue to the gauging bay. It is easy to see the periodic increase in queue contents throughout the system caused by an operator only being available at the load and unload bays during the first two shifts of each day, even though raw material arrives during all three shifts.

Interestingly, the buffer storage (BUF1) is not well utilised. It appears to be mainly responsible for an attempt at smoothing out the workflow through the system. However, it seems that its contents at each third shift may be increasing. This could imply a progressively intensive blockage is building up within the FMS, and certainly would warrant further investigation. Ultimately, the store is most busy at the start of each day, when workstation utilisations are necessarily 100 per cent, as indicated by the fact that their buffer queues (for example, MC1) are full to capacity at these times. The overall effect of this on the gauging machine is quite substantial. For a significant proportion of the simulated time, the queue in front of this equipment (GAGE) is full. This is especially the case at the beginning of each day, when there is an additional backlog of work to be cleared.

Turning to the pie charts shown in figure 7.3, it is clear that the FMS is not well balanced. MC1 is idle for nearly half the time, while the gauging equipment is blocked — that is, its input queue is full for much of the time. It would certainly be worth considering increasing the capacity of the queue in front of the this bay. Finally, referring to the system utilisation Gantt chart shown in figure 7.4, it is clear that the work flow through the system is far from smooth: it is too dependent on the operation of the load and unload bays.

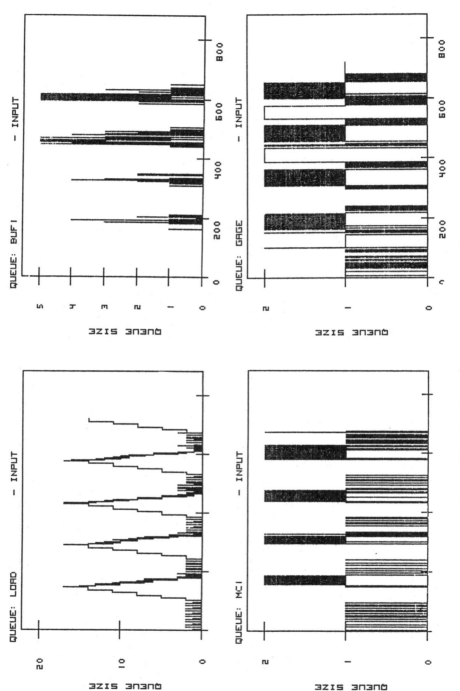

Figure 7.2   Graphics of queue content versus time

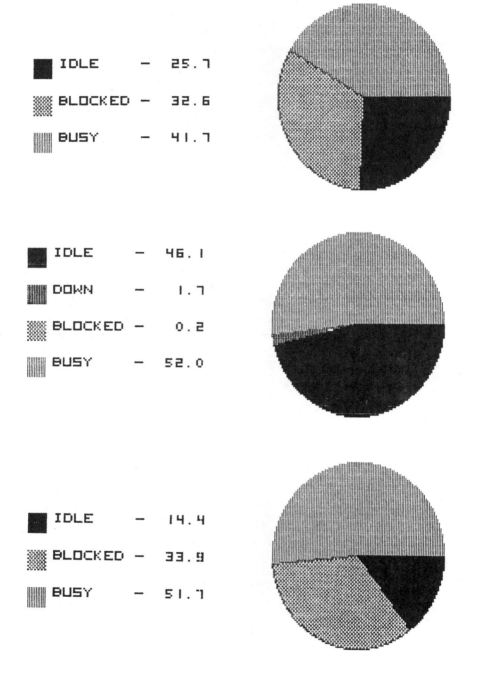

Figure 7.3 Pie charts of workstation utilisation

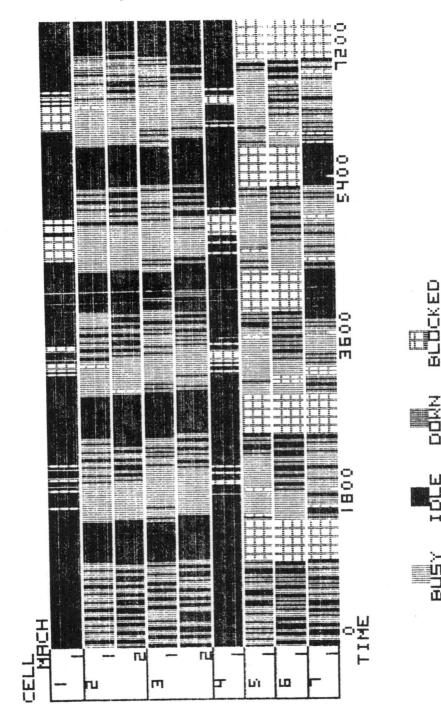

Figure 7.4 Gantt chart of system utilisation

In fact, many more conclusions can be drawn from this small simulation run and the few results which have been shown, though these will be left to the reader to identify.

Although the example given is relatively simple, hopefully it has nevertheless demonstrated not only that packages such as Modelmaster are extremely easy to use, but also that they can quickly generate extremely valuable information. Certainly, to create even this very simple model using a more traditional simulation package would represent a substantial task for a skilled individual; in fact it represents only a few hours' work.

7.5 How and when computer simulation should be used

There is no doubt that simulation should be used as much as possible throughout the implementation of an FMS. Indeed, there is no excuse for this not being the case when the simulation system is as easy to use as that just described.

As intimated earlier, there are a number of quite distinct phases through which the design of an FMS evolves before it changes from an idea into a reality. Simulation is able to assist with virtually all of these stages.

During the Conceptual Design phase, a simulation can add credibility to an idea. Although the design of the system will be far from fixed, there is no reason whey the model cannot be used to give an indication of the throughput of the system given a certain configuration of workstations. Such a model may also be used to great advantage when calculating the initial financial analysis, not to mention the fact that a well-thought-out animated display, as part of a Main-Board proposal, is likely to add considerable weight and credibility to the idea, and the work carried out so far.

During the Detailed Design phase in particular, simulation should be used continually, especially during the earlier stages. If a suggestion is made as to how to improve the system, the impact of the change should first be investigated using the model. During the later stages of the Detailed Design, many of the purchase decisions on major capital items will have been made, and it is therefore important to ensure that as much simulation as possible is carried before these commitments are made. It is certainly far better to be accused of over-simulating a system at an early stage of the project, rather than to find, after the system has been installed, that it does not work as expected.

Together, these two phases are probably where simulation is in some respects most helpful, essentially because the design is still fluid.

When the design is about to be stabilised it could well be appropriate, perhaps because of the limitations of even the best of microcomputer simulation systems, that a more sophisticated model is justified. This would not only assist with the analysis being carried out during the final phase of the Detailed Design, but would also confirm the results of the previous simulation effort. For example, results between the two aproaches could be compared. Also, the mainframe model, once created, should be able to simulate different working situations, in greater detail, more quickly than might have been possible with the microcomputer simulation.

For many Project Teams this will be an appropriate time to obtain some outside assistance, possibly from a university or perhaps a professional bureau simulation service. Some of the microcomputer modelling systems also have more sophisticated versions which run on mainframe computers. If this is known in advance, such a facility could greatly enhance the modelling process, since the original model could be transferred to a more powerful computer — mini or mainframe — and possibly then enhanced. However, while this might appear attractive, possibly as both a time and money saver, there is the risk that any errors in the smaller model might be transferred to the larger model. If a second simulation is to be carried out, there is much to be gained from having the second model created in an entirely different environment to that of the first simulation.

Of course, regardless of the type of simulation system being used, or who is carrying out the work, it is essential that a sufficiently large number of runs is made, ideally with variations in one factor only, to ensure that a statistically significant result is obtained. Obviously if this is not done, little confidence can be attributed to conclusions reached as a direct result of the simulation exercise.

For example, if the utilisation of, say, operators or AGVs is being considered with a view to making a decision as to how many will be required to ensure a particular rate of production, a typical procedure might be: fix the overall system design, develop maybe ten different work schedules (ideally reflecting the best and worst workload mixes), and then run the model with all these schedules with different numbers of operators or AGVs etc. The output of these runs could then be tabulated, as shown in figure 5.2, and the impact of the variable changes studied. However, a number of runs would also be needed where the collective impact of a number of simultaneously variable changes is also considered. This will provide knowledge about the relationships between the variables and how, together, they impact on overall system performance. Again, the ability of the modelling system to perform multiple runs automatically, given a number of pre-defined input files, greatly facilitates this type of simulation analysis, since, for example, the model may then be left to run overnight.

However, having completed the Detailed Design phase of the FMS implementation there is no reason why simulation investigations should not continue. The runs are likely to be rather less frequent since, hopefully, the design of the system has been fixed, though small modifications in the design are likely to occur, and the impact of these on the system as a whole should be well investigated, preferably by simulation, prior to their inclusion into the FMS design.

During the installation and commissioning phases of the FMS, simulation will not play a particularly significant role. But when the system is essentially commissioned and only debugging and production are required, simulation once again becomes an invaluable aid.

As a debugging tool, the simulation model may be used to test a particular set of operating circumstances probably far more quickly than would be possible within the FMS itself. Then, the same set of conditions may be created within the system, and if the results of the two tests correlate well, added credibility will be given to both the model and the FMS. If this correlation persists throughout many such tests, the model may well have proven itself as a valuable diagnostics tool and operational aid. If there is little or no correlation, an investigation would be justified. It might be appropriate to question the design of the FMS, or whether the system is actually operating as expected.

These two last uses of the simulation model are often ignored by the FMS designers, perhaps rather surprisingly. After all, having spent so much time and effort developing an accurate model of the FMS, there is no reason why it should not be used to help with the actual running of the system. For example, the effect of different work schedules or mixes of components can be simulated on the model before they are actually forced into the system. This gives the eventual System Manager the opportunity to study the impact of the schedule on, for example, machine utilisation and throughput time.

Diagnosis of faults within an FMS is frequently an extremely difficult task. Usually the fault, for example, a pallet appearing in the wrong place at the wrong time, or an incorrect tool having been used by a machine, is only apparent 'after the event'. The system operators perhaps see the symptom rather than the disease. Simulation can be of assistance in identifying the causes of a problem. Though, admittedly, what is required of the model to facilitate this diagnosis procedure is the ability to run the model 'backwards'. This is, of course, not a straightforward process and is perhaps more in the realms of expert systems (which are discussed in chapter 12). Nevertheless, simulation may still be used to advantage with the diagnosis of many complex faults occurring within an FMS, not to mention the more obvious problems such as bottleneck identification, and an analysis of the effect of feasible alternatives etc. Generally speaking, the level of simulation activity could be expected to vary during the implementation of the FMS, as shown in figure 7.5.

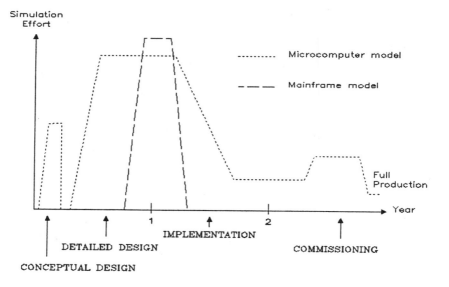

Figure 7.5   Typical levels of simulation usage during FMS projects

7.6   Concluding comments

There are few features of an FMS design analysis to which simulation cannot be turned to advantage. Typically:

- Component throughput
- Number of machine-tools
- Machine-tool utilisation
- Fixture requirements
- Tooling requirements
- Material transport speed, pallet loadings etc.
- Material transport, number of vehicles
- Number of operators required
- Operator utilisation
- Resilience to breakdown
- Effect of differing work schedules
- As an aid to financial justification
- Facility layout
- Floor-space requirements
- Work in progress
- Elimination of bottlenecks
- Developing system control algorithms

However, the above relatively small subset of FMS features which computer simulation could be used to investigate is certainly invaluable, and, perhaps in part at least explains why simulation continues to become more popular. There are also a number of other uses to which simulation may be put to advantage, not necessarily associated with the analysis of proposed manufacturing systems, such as:

- 'What if' analyses.
- As a means of identifying ways of increasing the efficiency of existing plant.
- As a means of comparing system alterations' alternatives.
- As a means of helping to integrate different systems. For example, the impact of establishing an FMS within an otherwise traditional existing manufacturing plant is frequently underestimated (if it is considered at all!).

Simulation is able to assist considerably with such analyses.

Finally, little has been said, so far, about the use of analytical techniques for assisting with the design of FMS installations. There are a number of such techniques available, probably the most popular being a mean value analysis. This is essentially a tool to study such factors as throughput, utilisation and mean queue length within a network of queues. Another popular analysis tool is the IDEF modelling technique. This is especially valuable for studying material and information flows.

Probably nobody would argue that analytical techniques are a complete substitute for computer simulation, especially at the present time. But in certain circumstances they can be extremely useful, for example, in testing the simulation itself, and while formulating major design principles.

An analytical technique will usually have to average out detail if it is not to become an extremely lengthy and costly exercise. After all, it is difficult enough trying to simulate a real factory situation, never mind trying to explain it in the form of analytical relationships. But certainly, if one is familiar with these techniques, they can be used to great advantage. Their application has been made much more simple by the availability of a number of microcomputer programs to take the considerable burden of manual calculation away from the system designer. However, with the great steps forward that are continuing to be made with microcomputer-based simulation packages, it seems that most requirements can and will be met by these.

8 Computer Control Systems

8.1 Introduction

All flexible manufacturing systems may be subdivided into three major components:

(1) Production equipment (process and material handling).
(2) A communications network.
(3) A computer control system.

As far as FMS is concerned, the progress that has been made in the development of production equipment, for example, machine-tools, robots etc., has been substantial, although the same cannot be said for either the communications or the computer systems necessary. In many respects these latter two are the most important and also the most complex aspects of any highly automated system. Possibly, this complexity has been responsible for the somewhat disappointing rate at which advances appear to have been made in this area. Certainly, significant progress has been made with regard to the computer hardware but the same is most definitely not the case where the control software is concerned.

The main purpose of this chapter is to discuss both the computer hardware and software required to control a typical FMS. Bearing in mind the obvious complexity of this topic, it should be appreciated that, at best, this chapter will only be able to serve as a generalised introduction to the subject. The following chapter will consider the issue of communications, while subjects such as reusable software and generic FMS control system software are discussed in chapter 12.

8.2 Some background

In many ways the computer control system represents the largest single risk element within the FMS. In terms of cost, the computer hardware and its associated software could easily account for a substantial proportion of the overall cost of the FMS. Typically, some 15–30 per cent of the FMS cost will be devoted to the provision of control software, with another 10–15 per cent being needed for the computer hardware. An FMS even with a good overall systems design might well be doomed to eventual failure if the computer control system does not function adequately. It is therefore not unreasonable to suggest that some 50–75 per cent of the total risk involved in implementing the FMS is represented by the development of the computer control system, which of course includes both the software and all the hardware on which it runs, much of which might not be standard.

This is certainly a sobering fact, and perhaps surprisingly an element to which many FMS designers pay inadequate attention. The reasons for this are probably many, but indubitably the most common is the fact that most flexible manufacturing systems have their 'mechanics' designed by machine-tool-oriented production engineers. However, the software is frequently designed by people who do not fully appreciate the intricacies of the manufacturing environment that they are about to try and control. This is to be expected given that these people are selected for their software skills, not their production engineering skills. If such a polarisation of expertise is allowed to evolve, it is hardly surprising that some flexible manufacturing systems do not perform as well as was anticipated.

This problem is further compounded by the fact that software is very difficult to appraise in terms of its ultimate capability prior to being installed in its entirety and tested on the FMS. Of course at this point, if the software, or hardware, is deemed not to be performing adequately, rectification is likely to be an expensive and embarrassing operation.

Because of the complex, multi-vendor nature of the computer control system, in terms of the fact that all equipment and activities occurring within the FMS will ultimately have to be monitored and controlled by the software developed, it is imperative, from the very outset of the project, that the penalties associated with an inadequate computer control system are fully understood by all suppliers of equipment and services to the FMS. Usually, the development of an inadequate control system results from the influence of two factors. First, badly written software or inappropriate hardware (the latter being slightly easier to rectify, but still expensive). Second, and probably most seriously, incorrectly specified software. Regardless of how well the software has been programmed, if it was not orginally specified correctly, it is almost inevitably not going to be able to perform the task of running the FMS correctly.

The message that the above is intended to convey is that the computer control system is a single factor that is capable of making or breaking the FMS. To ensure the latter does not occur, one needs to adhere to the following three rules:

(1) As far as possible, ensure that those who will be responsible for the coding of the software are highly skilled, capable and familiar with the type of task that will be expected of them.

Needless to say, it is extremely difficult to judge the competence of a team of systems analysts and programmers. This situation is not helped by the notoriously high inter-company transfer rate which typically occurs with these skilled people, a fact which itself confuses the issue of how practical it really is to attempt to ensure that a reasonable level of long-term support is maintained. Indeed one might as well resign oneself to the fact that the probability of the software team remaining available to provide support for any length of time after the notional completion of the software is extremely low. Therefore the prime concern must be to ensure that the software will be written and documented as well as possible. For this to be achieved, it is desirable that as many members of the software team as possible are experienced in the generation of real-time control systems, and preferably have been involved with the control and co-ordination of the type of manufacturing equipment to be employed within the FMS. The ideal situation is, of course, that several members of the software team have recently been associated with a similar, hopefully successful project.

(2) The software specification must be made comprehensive and fixed completely as soon as possible. Where appropriate, the implications of the software specification should also be identified and fixed. This might include features such as which equipment has plugs and which has sockets, and similarly with pin allocations and workstation controller push button sequences etc. While it is essential to have the software specification fixed, it is at least as important to establish the implications of this document and make them known to all the relevant parties.

Similar to the overall design of the FMS itself, a software specification is usually generated in three distinct phases. First, the Conceptual Design (which is sometimes called the Systems Definition). Second, the Functional Design, and finally, the Detailed Design. After this third phase has been completed, all that remains is to allocate areas of responsibility to individual programmers, and to schedule the generation and the testing of the software itself. Appropriate benchmarks should also be established in order that progress can be readily monitored. Obviously, great care should be taken to ensure that the software is generated in a manner which optimises a number of factors, such as the testing of the software in a modular and collective manner, the testing of the FMS in a modular and collective manner, and the debugging and start-up of the FMS.

As with the FMS itself, the Conceptual Design phase should be used to identify the major elements of the system, in this case software modules, how they will interact with each other and most importantly, how they will interact with the FMS itself, not only in terms of the equipment involved but also the operators, since more often than not it is these people who will have to interface most closely with the software. Additionally, it is not unusual for some estimates to be made of the type and size of computer hardware likely to be necessary to run the software.

During the Functional Design phase, considerable substance is added to the work previously carried out. Such factors as the user-interface would be considered in depth, together with the breakdown of the software into appropriately sized and definable tasks. Once the Functional Design of the software has been fixed, most aspects of the way in which the FMS is intended to operate are also fixed. It is important that all equipment suppliers are completely familiar with the implications of this document.

There should be no doubt in the minds of the designers that the simulation of the FMS and the design of the control software are among the most important aspects of the design. Failure to carry out these tasks correctly will result in the failure of the FMS. It is worthwhile emphasising this point, since it is these aspects of system design which are expensive without apparently generating anything of overt value (apart from lengthy reports which few may take time to read, and even fewer may understand). In both cases, key members of the Project Team are likely to be working in a quiet room, using up the project budget, at a time when upper management frequently would prefer to be able to relate to a more tangible, or perhaps more easily understood and visible, measure of progress, such as the arrival of a machine-tool. Great patience is required at this time. These formative stages of the FMS must not be rushed, they must instead be well considered and absolutely comprehensive.

(3) The final major factor to be given careful consideration is that of selecting the computer hardware on which the control system will eventually be run and maybe developed.

With the considerable advances that have been made with the development of computers, the selection of computer hardware should have become somewhat easier. Unfortunately, as is frequently the case, this has not really proven to be the case since although the traditional questions of, for example, performance are generally more easily answered, there are a number of other questions which also need to be answered. A good example of this might be the long-term implications of selecting a particular computer vendor, since not only the issue of long-term survival/support should be considered, but also how the hardware will integrate with other computer hardware systems currently installed, or due to be installed in the factory and/or company. With the power and reliability of computers continuing to increase while their unit sizes and environmental control requirements continue to decrease, the ability of a particular type of hardware to perform a task is without doubt becoming easier to assess. Even the issue of sizing has become somewhat easier, since software is generally becoming more portable and many computer systems are readily upgradable to more powerful units.

As discussed above, it is now really the secondary consequences of selecting a particular type of hardware which need to be investigated thoroughly, including such factors as long-term support (for both parts and labour) and the availability of software expertise (that is, language familiarity), development tools etc. However, this does not mean that the task of correctly sizing the computer's main memory, disk storage and response times should be underestimated in their importance.

If the influence of the above three itemised factors is considered in detail, the chances of the control system being well suited to the FMS are increased substantially. Much of the remainder of this chapter will be devoted to further discussion of the implications of these topics.

8.3   Objectives of the FMS control system

In general terms, the main objectives of an FMS computer control system are relatively straightforward to identify. Unfortunately, it is not quite so easy to develop the software necessary to meet these objectives. Essentially, the objectives are:

(1) To facilitate the transmission of support software (such as part programs etc.) to the material handling and process equipment within the FMS.
(2) Co-ordination of the activities of the equipment within the FMS. This ensures that the various material handling systems provide all that is necessary for the process equipment to function at as high a level of utilisation as is deemed desirable (for example, raw material, tools, jigs and fixtures etc.).
(3) To facilitate data entry, control, operation and monitoring of the FMS as a whole by the operators.
(4) To assist with the return of the system to a fully operational condition after a failure has occurred; currently this task is unlikely to be fully automated, and some manual intervention is almost always inevitable.

These objectives are virtually the same for all flexible manufacturing systems. As the chapter progresses, it will describe in more detail the attributes of the software necessary to control an FMS typical of that which might be implemented today in any discrete parts environment, largely because this is the area in which most FMS applications currently reside. However, every effort will be made to ensure that the descriptions remain appropriate to as wide a variety of automation systems as is practicable.

At the functional level, the FMS control software has to be capable of facilitating a number of activities within the flexible manufacturing facility. Most of these, which are apparent to the operators, are associated with ensuring that the correct data is available to the computer, thus enabling it to continue to control the FMS. This information needs to be in a form which is easily updated to reflect changes in the operating environment. The data is essentially as listed below:

- Component and process information
- Part programs and set-up information
- Ancillary process data (tool wear rates etc.)
- Calendar and capacity information
- Order information
- Work scheduling algorithms
- Communications handling
- Contingency management
- Status monitoring
- Standby operation and data back-up
- Statistics
- Remote diagnostics
- Co-ordination with other plant activities

However, prior to describing the typical requirements of the software supporting the maintenance of the above data, it is appropriate to discuss briefly the user-interface to the software. This is essentially the content and layout of the screens which the operators see when they use the computer system and its peripherals, and the type of response required to allow the computer system to proceed. Obviously the user-interface to any computer system should be designed to be as simple and as pleasurable to use as possible. There has been much written about the need to have more 'user-friendly' software, and probably there is little that can be added at this point. Having said this however, there are some features of FMS software which make its requirements unique. For example, the system is likely to be used by people of a variety of different skill levels, many possibly being computer novices. Therefore 'help' comments should always be readily available, perhaps by the pressing of a designated 'help' key which would always display information pertinent to the operations currently being carried out (that is, it is context-sensitive). Having different levels of assistance readily available to accommodate different skill/familiarity levels would naturally be an added benefit. However, it is likely that once the system has been fully debugged and used for a suitable acceptance period, the help functions will not be needed to such an extent.

Following on from this, every effort should be made to make the interface easy to use by everyone, whether they are a highly skilled System Manager, or an apprentice operator who is barely familiar with a computer terminal keyboard. The most common way of achieving this is by using a menu. Items are displayed in a list, usually numbered, and selections can be made either by typing and entering the appropriate number, or perhaps by moving the cursor

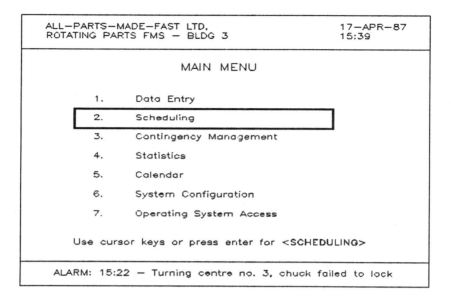

Figure 8.1   Typical FMS menu layout

keys up and down (in which case it is usual for the selected item to be highlighted in reverse video). Usually such a system proves both easy and fast to use, especially for an experienced operator. A typical display might be as shown in figure 8.1, which also has areas on the screen for messages generated by the system, the time etc.

The menus should be organised into a simple-to-understand tree structure; this will facilitate travelling from one function to another. Also they should all be created using the same format, so that whatever the function being performed, the way in which the menu is used will be the same. The careful use of colour, function keys, protected fields to facilitate data input and checking, the ability to transfer from one menu to another and the incorporation of light pen, 'mouse' or cross-hair operation are all features which can considerably improve the user-friendliness of any software package.

The ability of the user-interface to 'make or break' a software package should not be underestimated. If a software package is difficult to use, it will become unpopular surprisingly quickly, regardless of how efficiently the package actually operates. If software becomes unpopular, people tend not to use it properly, and then perhaps management deem the software to be inadequate, not because it is technically incapable of carrying out the intended task, but because nobody wants to use it. (This has been the fate of several production control packages which proved so difficult for the operators to use that data was not input regularly. With the data inaccurate, the software was doomed to failure.)

8.4 General software requirements

The FMS computer system needs to have input, and to be able to retain all the information relating to the parts which will be needed to control their manufacture. Of course, unless the software has been appropriately specified, subtle differences between the part's physical manufacturing requirements will not be apparent to the FMS control system, which will simply be checking the completeness of the data being entered. The type of information which will have to be available to the computer system, is as follows:

- Part number
- Part description
- Raw material type and value
- Part type
- Finished part value
- Load and unload instructions
- Palletisation details
- Pallet fixturing requirements
- Normal production process sequence
- Alternative process sequence

Typical operations which would be carried out on this information would be addition of further data, together with the modification and deletion of existing data.

For a part to be produced within a highly automated manufacturing system, some means of identifying the part must be available for both the computer system and the operators. The above information would be associated with one particular element, typically the part number. Obviously, the number of such descriptions will depend on the way in which the manufacturing system is deemed to be flexible. If, for example, the FMS is within a near mass production environment, there might be only six or ten such part descriptions. If it is in a jobbing shop environment, there could be hundreds or even thousands.

With the above data entered, it is then necessary to define the processes which will be used to manufacture the part. In certain circumstances this information might well be combined with the part definition. This would typically be the case within a manufacturing system where the production processes were finished-part-dependent. If the production processes could be used for a significantly large variety of different parts, it might prove more efficient to define the processes in such a way that they are independent of the part. The information which would be required for this function would be:

- Operation number
- Process name
- Workstation or bay location
- Part program(s) required
- Part capacity per cycle
- Number of parts 'in' and 'out' per cycle
- Set-up file(s) required
- Tools, jigs and fixtures required
- Tool offsets and wear characteristics
- Inspection requirements
- Ancillary inspection-related data
- Typical process cycle time
- Acceptable cycle time tolerance

As with all such data available to the computer system, it will usually prove necessary to provide the facility to add, delete and modify these process definitions. It might also be necessary to define whether or not any process-related buffer stores are required etc., depending on the configuration of the FMS. At this point the descriptive information necessary to manufacture a part has at least been defined if not input to the system. All that remains is to provide the facility to input the necessary process-dependent files which the actual workstation will use to manufacture the parts.

The type of files which would be needed are machine-tool, robot part programs or programmable controller programs and recipe files. One of the key issues with this type of data is that of ensuring that the correct version of any particular program is always being used. This will probably require a degree of operator discipline, but certainly the control software should be designed to assist with this whenever possible. For example, it might be decided that the version of the part program held on the host computer is the master. As such it might have a name which reflects the finished part and process/operation with which it is also associated. The version number will reflect the creation date, for example, 'PARTNAME.OPN;3' could be the program for operation 'N' on the part 'PARTNAME', and the latest version of the program is number '3'. If a new version was received by the computer system, it would become version '4', and would become the version used for production, with version '3' eventually being deleted. If an unproven program arrives from an associated CAD system, it might prove desirable to ensure that the name by which this program is temporarily stored is not that of a production program. One possibility is to add a prefix to the program name which reflects the place of origination (for example, CADPARTNAME.OPN;5 for a program developed on a CAD system etc.). Various ancillary files (such as tool offset tables, robot parameter tables etc.) would need to be treated in a similar manner.

Set-up files would probably also be needed. Typically, these are text files, possibly interactive, which are designed to assist an operator with the setting-up or stripping-down of a workstation either before or after having produced an order of particular parts. Files similar to these could well be used as part contingency management, namely the process of returning equipment to normal working after it has failed.

At this point, the computer system has available all the information necessary to manufacture a particular part. What is needed now is information relating to the capacity of the system. This information comprises two parts. First, there is the number of workstations within the FMS. This might be the number of machine tools and robots (in a machining FMS), or the number of assembly or disassembly stations (in an assembly or disassembly FMS) etc. Of course, it might not be a fixed number, since that capacity of a system which is not fully automated will depend on the number of operators available during a particular shift.

Secondly, the capacity of the FMS will obviously depend on the number of hours during which it is to be operated. If the system is to be operated all year, seven days a week, then obviously there is a possibility that the system will be operational for 24 hours on all 365 days. This time would typically be subdivided into three eight-hours shifts on each day. However, the system is unlikely to be operated throughout the entire year, for example, there will be holidays and perhaps preventative maintenance periods to be taken into account. This information would need to be entered into the FMS operational time file or what is often called the 'calendar file'. Some sections of this file would be updated only infrequently, to reflect changes in annual working times etc. Similar to the other functions mentioned, the facility to modify this calendar file would need to exist. The ease of use of this facility could reflect the frequency at which it is likely to be used. For example, if the file is only likely to be updated once each year, this task might well be carried out by the System Manager. If the file is to be updated at the start of each shift as people 'clock-in' to the facility, then a more polished user-friendly environment would be appropriate. However, some of the data might well be updated on an hourly basis, to reflect the availability of operators both when they are expected to become available and when they actually do become available.

The next requirement is that of entering an order for a batch of parts. Since by this stage the system knows what parts it is able to accommodate, the workstations and their associated capacity constraints, the only additional information required by the computer system is essentially as follows:

- Order number or name
- Order description
- Order quantity
- Date due
- Special processing instructions

- Special loading instructions
- Special unload instructions
- Order priority

Leaving the creation of orders until last in terms of data entry assumes that this part of the data within the FMS is likely to change most frequently. If this were not the case, it could perhaps be more appropriate to enter the data in a different sequence. However, it is likely that the overall data requirements will be essentially the same regardless of the particular FMS application.

At this point, the computer system should have all the information needed to control the manufacture of the parts, and prior to committing the order to the system it would be appropriate to check that this information is indeed available, and to issue an error message if it is not. However, in certain circumstances it might be deemed desirable to allow an order to be committed without all the necessary production data being available (perhaps it is known that it will be available shortly), therefore it is likely that a manual override facility will need to be provided.

As an aside, it is often useful to provide some trace as to when and by whom all the relevant data was input. Since the input time and date are always known to the system, this information is easily included with appropriate data. However, regarding the source, it would be necessary to ask an operator to input an identification code. It is extremely useful if each operator is allocated such a code. This not only facilitates the tracing of data origination, but also the integrity of data made available, for example, for time-keeping. This identification code could also be used as the basis of a password protection scheme which could make different levels of software functionality available to different operators, such as denying direct operating system access to all but the System Manager etc. However, it is probably not desirable that each operator should always have to log-on to the system whenever an interaction is required. This would probably introduce an unacceptable level of tedium into the system, not to mention raising the issue of failures to log-off. To overcome this, therefore, a default level of functionality should be available to all operators of the system.

8.5 Work scheduling

With all this information, together with that entered previously, the next task is for either the system or one of the operators to decide which order of parts should be processed within the FMS at any particular time. This means creating a work schedule for the FMS, a complex task and yet one which is carried out frequently and impacts significantly on the eventual productivity of the FMS. The FMS computer system is unlikely to be able to carry out this task without any manual intervention, unless it has been equipped with what to date, at least, has proved to be somewhat difficult to realise — a fully automatic work scheduling facility. So usually a compromise solution is provided, with an automatic facility performing some of the work, while the final schedule is created manually. Usually, the automatic portion will incorporate some algorithms specific to the type of manufacturing environment in which the FMS is installed. These could perhaps be tuned, to take account of variations in the work scheduling requirements.

Ultimately, the requirements of an FMS work scheduling system are to take the workload (that is, the orders to be manufactured) and, within the constraints applicable to the FMS (that is, workstations, operators, working hours etc.) schedule the various orders to occupy available resources in a particular time slot (including allowances and responses to unplanned maintenance) which optimises one or more constraints, such as workstation utilisation, work in progress, order lateness etc. Obviously, this is not an easy task, and yet it is one which is of fundamental importance to the success of the FMS.

Much has been written about work scheduling especially since the 1950s, as applied to job shops, the 'general case' and flexible manufacturing systems. As yet, there do not appear to be any FMS installations which demonstrate truly automatic work scheduling, especially not with any schedule optimisation included. No doubt it is only a matter of time before one is developed but, in the meantime, there are a number of ways in which the work scheduling problem can be sufficiently simplified to become solvable to an acceptable degree.

Of course, the real difficulty associated with scheduling, say n jobs through p processes is the combinatorial nature of the problem, or to use the correct jargon, that the solution is 'np hard'. Given that the optimisation of even a relatively small scheduling problem, manually, is likely to be an impossible task, the way in which most traditional job shops solve this problem is with the aid of a Gantt chart. This is essentially a bar chart with the length of bars denoting the duration of the requirement of a resource for a particular task. An example of such a chart from the SCAMP FMS is shown in figure 8.2.

This procedure, which is obviously closely related to that of the PERT diagram and Critical Path Analysis, has been applied widely in such diverse fields as computer, taxicab and police car allocation. It is a uniquely effective means of planning the allocation of scarce resources. However, this type of system remains a means of simplifying the problem of allocating resources, albeit with the help of a computer display. The Gantt chart itself does not help much with the optimising of the work schedule. For this an entirely different approach is needed, though Gantt charts will no doubt remain one of the most popular ways to display works schedules, and perhaps form the basis, on computer displays, for manual modification.

Having discussed in general the complexities of creating and displaying a work schedule for a flexible manufacturing system, the problem of how to specify the software needed to carry out this task remains. Unfortunately,

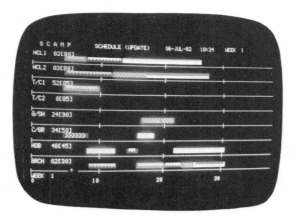

*Figure 8.2   A typical SCAMP work schedule Gantt chart
(courtesy of SCAMP Systems Ltd)*

there is no easy solution to this problem. A fundamental requirement, however, is a thorough understanding of the scheduling requirements, and constraints, of the FMS being designed. Ultimately, the FMS comprises a number of resources and constraints. The FMS control software, by checking the availability and allocating these resources in a particular manner, should be able to effect some measure of the productivity of the FMS. This measure should be checked for feasibility and acceptability prior to being used by the FMS. Almost inevitably, therefore, the scheduling software will either have to use an algorithm to produce a sequence in which to carry out the jobs or some form of iterative process will be needed, if not a combination of the two.

As was discussed briefly in chapter 7, an interesting and powerful addition to FMS control software is the ability to run, in parallel, a simulation of the FMS and forecast how the FMS will react to a certain set of circumstances. A facility such as this has a surprisingly wide influence on the way in which FMS is operated, since not only can it be used to assist with the scheduling of work for the FMS but also, using the system's status displays, it may be used as a means of helping with fault diagnosis and equipment restart, which will be discussed in the following section.

Having created a work schedule, the provision for manual modification must be provided. This requirement exists not only because the operators will occasionally wish to adjust or fine-tune the schedule, perhaps in response to a change in the overall production requirement, but also because progress within the FMS is unlikely to have been as planned. For example, orders will not be completed when scheduled, workstations will break down, tools will fail, some jobs will even finish early, etc. The FMS scheduling system needs to be able to respond quickly to these changes and to develop at least a feasible course of action in all forseeable instances. For this to occur, factors such as the maximum allowable overlap between various operations — that is, different pallets from one batch being available to different operations, expected transport times etc. — will all need to be taken into account.

The time needed to generate a new work schedule might itself create a problem. Since the schedule reflects the real-time status of the FMS, generating a new schedule to accommodate some change in state means that by the time the new schedule has been created, the state might well have changed again. This is particularly likely to be the case if the new schedule is being created manually. Therefore attention must be given to ensuring that any new schedules are brought up-to-date, or synchronised, with the new status of the FMS.

Once a work schedule has been created for the FMS the next task is to issue relevant instructions to the various workstations, in order actually to initiate the manufacturing processes. This comprises essentially two activities:

(1) Communicating the elements of the particular task to the appropriate equipment. (This is a subject which will be addressed in chapter 9.)
(2) Subdividing the overall task into manageable elements with start and finish times.

The second of these is a particularly complex operation and one which is fundamental to the successful operation of the FMS. All the activities which occur within the FMS will need in some way to be monitored by the FMS control software. A substantial proportion of these activities will be driven by the content of the work schedule, the exceptions being contingency management requirements. Therefore every activity will need to be broken down into individual functions, with appropriate interactions being clearly specified. The influence of this task spreads far beyond that of simply ensuring that work schedules are manageable. It also impacts considerably on how the computer system selects which particular software module to run next. Just as the resources within the FMS must be used optimally, so must those of the computer system itself — this, ultimately, being an issue of overall software architecture, a topic which is beyond the scope of this text.

8.6 Status monitoring and control

The status monitoring of flexible manufacturing systems and indeed any complex automated system comprises two tasks. First, the obtaining of the desired data and second, its presentation. The data is likely to consist of a number of different types of information, for example, information relating to the status of equipment, or the location of operators and equipment. This type of data is relatively straightforward to acquire provided that the necessary requirements have been specified early enough, and that correspondingly sophisticated hardware exists at the various controllers to extract the data and then communicate it to the host computer system. For example, data such as:

- Workstation on, off, busy or idle
- Optional stops enabled or disabled
- Workstation waiting for workstation
- Workstation in manual or auto mode
- Workstation requires assistance
- Workstation in alarm state

Many of these status messages (which represent only a small proportion of those typically required) would be accompanied by some supplementary information. For example, at the most general message level there would probably be a number of 'workstation requires assistance' and 'workstation in alarm' messages. If the operator required further information, the message could be augmented by the following:

'hydraulics fault'
'check fault'
'controller fault
'coolant fault' etc.

This type of information can generally be regarded as 'hard' because it is generated internal to the system and, assuming all the relevant systems are functioning reliably, the information is likely to be accurate.

Within the FMS there will also probably be a requirement to collect some 'soft' data. This data is operator-generated, either at a computer terminal, at a workstation controller, or perhaps at a workstation interface unit. This information is probably correct, but by its nature, this cannot be guaranteed, since an error could have been made during its entry or the operator might have been mistaken.

Certainly, the timely delivery of this information cannot be guaranteed. For example, in a relatively demanned environment there will be some delay before an operator moves to a certain location and inputs information into the FMS computer system. Such delays should be assumed to exist, and wherever possible the progress of the automated system should not be made dependent on the input of such 'soft' data. In certain circumstances, however, such as recovery, it is likely that some sections of the system will have to remain idle pending the input of information (such as whether a component which has been damaged in a machine is to be scrapped or not etc.) — a situation which might well exist if some or all of the parts have to be manually inspected before production can continue. It could prove desirable for quality tracking purposes to keep parts flowing through the system in a particular order. In this case, it is not unlikely that, at some stage, production will be held up pending the manual input of an inspection result.

Unfortunately, it is virtually impossible to eliminate circumstances such as these in a system, which means that on occasion the system must be reliant on operator input. However, it is worthwhile taking care, first, to minimise such situations, and second, where they are totally unavoidable, to make their occurrence as independent as possible of the operation of the rest of the system. This will help to ensure that if one part of the system has stopped, the remainder of the system is at least able to continue for a reasonable length of time: namely, that amount of time during which it is expected that the halted equipment will be restarted. How this data is treated within any one system will be dependent on the installation's requirements and its complexity.

The second aspect of status monitoring is the presentation of the data to the operators, and eventually upper management. There are a number of ways in which this could be done, but the most traditional is certainly that of printing the data out. This is useful in certain circumstances, for example, during the debugging phase, or as data to be used later. However, one of the opportunities offered by FMS is that of moving closer towards the goal of a paper-free factory. Therefore as much information as possible should be made available to computer terminals, and indeed shop-floor terminals, if appropriate. This might mean that the software will have to filter out unwanted information, and display the remainder in a succinct manner, which is easily comprehensible. For example, textural displays must be presented in an easy-to-understand manner. Certainly, all status messages should be decoded by the time an operator has to look at them. For example, if a particular workstation generates a status message type '36', the operator is really interested in knowing that this means there is a 'hydraulics fault' on the machine. Therefore this is the information which should be displayed, not '36'.

It is useful to have a console which is dedicated to printing out major status changes. A helpful feature is to be able to alter the level of filtering of status messages from different equipment, allowing operators to study carefully those sections of the FMS which might be failing, and not to have the relevant messages submerged by a vast quantity of irrelevant information. The console then provides a continuous record of what has and is happening in the system, including the computers themselves. In the event of a complex system fault developing, such a log can prove to be invaluable, albeit lengthy to analyse. However, this process can be simplified by having the log committed to disk and available for later message sorting and printing.

The use of a colour mimic display to show, if practicable, the status of the whole FMS is to be thoroughly recommended. Displays of this type are rapidly increasing in popularity as they become both more sophisticated and easier to implement. For example, it is not uncommon to have a graphical representation of an FMS displayed on a colour monitor with all the workstations clearly identified by 'icons' corresponding quite closely to the physical configuration of the equipment. The colour of the icons would then be changed to represent high-level status changes. Green might represent the

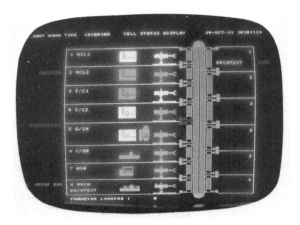

*Figure 8.3   An example of a SCAMP mimic display
(courtesy of SCAMP systems Ltd)*

equipment in the normal productive mode, while red might mean alarm, blue — requiring operator assistance, and so on. Such displays, while obviously most effective in colour, can also be made quite acceptable on standard black and white monitors. An example of typical such colour display from the SCAMP FMS is shown in figure 8.3.

In addition to displaying the overall FMS status, further information could be provided, perhaps by using a light pen to 'select' a particular item of equipment, or possibly by moving the screen cursor to the required equipment using either cross-hairs or just cursor control keys, depending on what the terminal has available. On selecting a particular workstation, data such as the order being processed, parts manufactured so far etc. could be displayed. Using techniques such as these, it is relatively straightforward to produce an interesting and effective method of data presentation. Such features will indubitably enhance the FMS as a whole, by making it easier to use.

8.7   Contingency management

Having input sufficient information to allow the FMS to be productive, and having also succeeded in displaying the precise status of the various items of equipment within the system, the next subject to consider is that of when an equipment failure occurs. Frequently this area of FMS control is called 'Contingency Management' or 'Recovery'. To put this subject in perspective, the software required to address this topic will represent between 50 and 80 per cent of the entire software required to control the FMS. This is essentially because, once something has gone wrong within a complex system such as an

FMS, it could be extremely difficult to return to a normal working condition, especially if the database of the computer system, which is meant to always reflect the true status of the FMS, is largely incorrect. If the FMS control software is unable to deal with the particular problem, it could even prove necessary, as a last resort, to empty the system completely of all parts, and then to restart the complete FMS. Such situations can be time-consuming, extremely expensive and damage credibility. It is therefore imperative that every conceivable failure mode of the FMS is predicted, and a detailed recovery procedure identified. Many failure modes will be subsets of others, but if feasible, all should first be considered as separate entities. Ultimately, contingency management procedures are likely to be a combination of both computer system activity and operator interaction.

It is difficult to recommend a particular method of approaching the definition of this somewhat application-specific problem and its solution. At some time or other, almost everything that could go wrong with the FMS probably will. However, having said this it usually transpires that faults may be subdivided into two main categories. First, those problems which are likely to occur relatively frequently and, as such, tend to be relatively straight-forward to trace. Second, those problems which occur less frequently and whose cause(s) are difficult to identify, and hence the action required to return the FMS to normal working is rather more complex to identify.

The first category might warrant actions such as:

- Increase the size of an order
- Decrease the size of an order
- Remove an order from the system
- Trace the components produced since the last manual inspection
- Change the contents of a pallet
- Change a pallet
- Empty a buffer store
- Test a communications line

Facilities such as these (and many more like them) should be readily available to operators via appropriate menus. Indeed if these functions are chosen well — that is, if they are both comprehensive and suitably generic — it should be possible to use a combination of these to recover from virtually any error state.

An example of a typical circumstance in which it might prove necessary to use such a recovery facility could be when a robot has dropped a component it was attempting to load into or unload from a machine-tool. A typical sequence of events which might occur as the problem arises could be:

- Machine-tool status, OK
- Robot status, idle
- Machine-tool status, waiting for robot

- Robot, busy
- Robot status, alarm grippers closed (this would be because a component had been dropped by the robot)

In this situation the machine-tool remains operational, however it is being delayed pending the correct delivery of a component. Meanwhile the robot has dropped the part (which may or may not be damaged) and has ceased to function pending some operator-generated input as to whether or not it is actually safe to continue. So the FMS mimic display picture on the system status display could perhaps be showing a green flashing machine icon (to indicate the OK but waiting status), while the robot icon would be red and flashing (to indicate to an operator that it requires urgent attention). Conceivably, a beacon on top of the robot could be illuminated to indicate the fault state, since such a device might attract the attention of an operator who is not near a mimic display. The operator would go to the robot and then possibly the following would occur:

- Place robot into unplanned maintenance mode
- Using robot diagnostics and visual inspection, ascertain cause of failure
- Disable robot (with correctly designed robot guarding, this would occur anyway as the operator entered the robot's working area).
- Enter robot working area and find component
- Check robot is not damaged
- Check component is not damaged
- Place component on original departure point
- Leave robot working area
- Advise host computer of current equipment and component status
- Reload robot parameters and/or program
- Return robot to operational mode
- Wait for host computer to enable cycle start
- Cycle start robot
- Robot icon on status display becomes green
- Robot successfully loads component into the machine
- Machine icon on status display returns to green

Obviously the above analysis is far from complete, as many of the questions asked have not been followed through. However, since the precise status changes and actions which need to be taken will probably vary substantially from one system to another, hopefully the above does at least give an indication of the amount of consideration which must be given to every such failure mode.

As was mentioned previously, it is highly likely that a combination of these various recovery functions will be needed to restart equipment in certain circumstances. The key issue is to ensure that these functions are as user-friendly as is reasonably possible, and that there are no circumstances from which, by using the correct combination of functions, it is not possible to recover.

8.8 Computer system back-up

One of the major problems whose effect should be minimised is that of a failure involving the controlling computer system itself. Protection against this type of failure is difficult. At the lowest level, adequate provision should be made for the back-up of data. This might be to a magnetic tape device, a back-up disk pack and/or drive or perhaps another computer system, possibly a larger system responsible for running the major part of the factory as opposed to just the FMS. The precise nature of the most appropriate back-up procedure will depend on the type of FMS being controlled and the overall architecture of the controlling computer system. In either case, user-friendliness is an important requirement. Ideally, the back-up procedure should implement itself automatically at regular intervals pre-set by the System Manager. The data archived should naturally include all part-related and process-related data, together with status data if this could be used to facilitate a restart of the FMS after, say, a head crash on one of the hard disks. One of the problems which frequently occurs with data back-up is that the process can be somewhat time-consuming, as far as the computer system is concerned. However, while the data is being archived it is likely that the FMS will still have to continue to operate. Ensuring that this is technically feasible can be quite difficult.

One of the most popular ways around this problem is to provide an alternative processor for the FMS control system. This essentially means providing the FMS with two computer systems each of which could individually control the FMS in isolation from the other. There are a number of variants on this theme, such as purely having dual disk storage systems, but the dual processor system appears to be becoming the most popular. Having said this, there are a number of ways in which the dual processor system could operate. For example, it could be a full 'hot' standby. In this case, when the computer system controlling the FMS fails for whatever reason, the back-up processor takes over the operation of the FMS automatically, and hopefully sufficiently quickly to ensure that the continuous operation of the workstations etc. is maintained. Then there is the possibility of 'warm' standby. Here the back-up processor is fully operational, but a certain amount of manual intervention is required to complete the changeover process of FMS control responsibility. Finally, there is the 'cold' standby. In this environment, the back-up processor might well be being utilised for an entirely different task, when required to control the FMS. In this instance, operators would be required to close-down these tasks and restart the computer system as the FMS controller. All these options have a number of advantages and disadvantages, the most significant of which are summarised as shown in figure 8.4.

Certainly if one can afford 'hot' stand-by, it is the most effective form of control back-up that exists. However, it is usually complex to implement in software terms and possibly in hardware terms as well, since all the FMS communications lines might need to be switched over to the back-up processor (and, of course, this could itself introduce a single point of failure). The advances which are currently being made in the development of 'cluster' computers might significantly reduce the cost of implementing a 'hot' standby computer system.

The most significant advantage that 'hot' standby offers is the absence of operator intervention in changing computers, and therefore there is a high degree of certainty that the FMS will continue to function after one computer system fails. If the FMS is to be operated 24 hours each day (and this is obviously highly desirable) then such a facility might well be the only viable option, and certainly it is a facility which is virtually impossible to retrofit. Having said this, it should be borne in mind that there is bound to be a short, albeit small, delay before the back-up processor system comes on-line. Care should be taken that any FMS status-change messages are not irretrievably lost when this occurs.

TYPE	CHANGEOVER TIME	OPERATOR ASSISTANCE	SOFTWARE COMPLEXITY	COST
HOT	V. short	None	High	V. high
COLD	Medium	Medium	Medium	Medium
WARM	Lengthy	V. high	Medium	Medium

Figure 8.4   Different approaches to standby computers

'Cold' and 'warm' standby configurations remain attractive in some environments, largely because of their reduced cost and complexity, provided:

(1) It is never intended to operate the FMS 24 hours each day.

(2) The FMS is able to operate for a reasonable amount of time without a computer system.

(3) It is acceptable that the FMS might be totally unproductive while the computer is repaired.

If this is the case, then these back-up options are well worth considering, especially where the mean time between FMS status changes is quite long (since then the number of status messages which might be lost is much reduced). Even if there is some doubt about the above, a 'hot' standby configuration should at least be investigated.

Regardless of the type of computer back-up provided for the FMS, a further facility which should be considered is that of providing a remote diagnostic capability. This permits, for example, the software developers, who may be located at a considerable distance from the FMS installation, to connect a terminal to the FMS as and when required. This could perhaps be implemented via a standard modem and a telephone line. Such a capability greatly facilitates debugging, especially after the FMS has been put into full production.

8.9 Statistics

The final major software area to be considered is that of statistics. While every effort should be made to ensure the operation of an FMS is as near paperless as possible, it is a fact of life that some statistics on the output etc. of the FMS will have to be generated. Although usually it might be acceptable to have these available at a computer terminal, at some stage, some operating statistics will almost inevitably have to be printed out. The type of statistics usually required are as follows:

- Equipment utilisation and uptime
- Mean time between failures and failure duration
- Order throughput times
- Cycle times per part
- Number of inspections/results
- Operator log-on/off times
- Communication faults
- Time spent developing part programs
- Buffer storage utilisation
- Equipment and operator status messages

These statistics would probably be required on a shift, daily, weekly, monthly, annual and cumulative basis, perhaps dependent on the data involved. Possibly, to obviate the need to archive significant amounts of data, these could be printed out automatically at intervals at the discretion of the System Manager.

8.10   General considerations

Early in the preparation of the design of the software, a decision will have to be made as to what type of computer hardware will be used. While the selection of an operating system and the language in which the software is to be programmed will have some bearing on this decision, it is becoming less important because of the increased portability of both. Nevertheless, the overall characterstics of the computer hardware and software combination comprising the FMS control system, which are of considerable interest to the eventual user, perhaps unlike those of the language used, are as follows:

- System performance must be adequate (for example, response to requests for assistance should be sufficiently fast to ensure that the overall FMS performance is not compromised).
- System development should not be made unnecessarily difficult (for example, a language and operating system where adequate experience and development tools are available should be selected).
- Long-term support should not be jeopardised (that is, a language and operating system which are likely to be difficult to support in the long term should not be selected).

Long-term maintenance is also an issue which should be considered carefully. When the FMS is in full production, if a problem develops, old or new, hardware or software-oriented, it is important that some expertise is available to solve the problem. Similar consideration should be given to any third-party hardware or software packages embedded in the FMS control system (such as database managers etc.), since these too, can be of fundamental importance to the long-term well-being of the FMS. Whether this support is provided by in-house or outside personnel will depend largely on the installation.

Following on from the above, a further maintenance-related issue is that of what response time should be guaranteed by the people responsible for maintenance in the event of a call-out being made. To some extent this will depend on the system configuration. For example, the same urgency might not exist in a system with 'hot' standby as with a system with no standby computer or with a remote diagnostics capability. Probably, a 24-hour call-out response is highly desirable. Due consideration should also be given as to whether or not some spares should be held on site for certain extremely important and/or specialised items of control hardware. Certainly, prevention is better than cure, and therefore adequate precautions should be taken to ensure that the system is not exposed to unnecessary abuse. This might, for example, mean denying the operators of the FMS direct access to the computer's operating system.

As far as hardware is concerned, terminals on the shopfloor, whether workstation interface units (possibly without workstations) or standard computer displays, should be plentiful. Certainly, there is nothing worse than trying to debug a complex problem and being a long way from the nearest computer communications device. Depending on the manufacturing environment, the terminals would need to be appropriately ruggedised to withstand oil spillages, airborne dust, electrical interference etc.

As computers become more reliable and able to withstand harsher operating conditions, the need for a carefully controlled environment in which to house the computers is rapidly disappearing. However, it is useful to have a central point from which all the FMS may be observed, especially as the facility is likely to be demanned. It is appropriate that this central point becomes the nerve centre of the FMS. It should provide an appropriate environment for the computer system and its ancillary equipment, together with room for the operators to observe the FMS. A colour mimic display within this room would obviously be advantageous.

Regarding the selection of computer system hardware, there are essentially three major considerations, all linked strongly to the design of the particular FMS. These are the processor main memory size, the response times required of the main processor and the disk storage capacity to be provided.

One of the advantages that has come from the considerable progress which has been made in the development of computers over the past few years, is that it is becoming relatively straightforward to upgrade processors and add disk capacity, thus making the actual selection process somewhat less critical, though this remains a relatively costly exercise. However, the possibility may well arise where the ideal hardware selection appears to lie at the top of a particular system's performance capability. If this is the case, then upwards expansion might not be possible, or at best might be prohibitively expensive. In this instance, it is even more important to take some time to clearly define the actual requirements of the computer system as a whole.

The selection of main memory size and the identification of response time requirements are to some degree linked. The first step is to estimate the size of the various modules likely to make up the control software, and from this, to estimate how much needs to be resident within the processor memory at any one time, taking into account the impact of the various techniques which may be used to ensure all the control software does not need to be installed simultaneously. There will probably be a significant difference between the start-up, close-down and steady-state running requirements, and naturally the worst case will have to be catered for, unless some drop in performance can be absorbed at these times. This could well be acceptable during start-up and close-down, especially if, for example, the system is expected to remain in operation for virtually all the time. But at times, when a recovery is required, or when rescheduling is in progress, the computer system has to be able to carry out these tasks quickly, and still be able to service the demands of the equipment within the FMS in a timely manner.

The level of response times needed will probably be dictated by a number of factors, for example, how often will it be necessary to poll each of the items of equipment within the FMS, once each second, or once each five or ten seconds? How much processing will be required following such a poll? What burden is placed on the processor by activities such as scheduling? How often will part-program uploads or downloads take place? Depending on answers to questions such as these, and there are considerably more which should be asked, an indication emerges as to the power required of the main processor. Also, a crucial influence on the overall system response time is exerted by the number of disk accesses which need to occur.

Also there are operator interactions to be considered. It can be extremely irritating, when trying to carry out a task quickly and after having entered the data, having to wait for what seems to be a totally unreasonable amount of time for such a simple task (possibly only a few seconds), while the computer is busy attending to another task.

The main requirement for disk storage will be to accommodate all the control software itself, to provide sufficient capacity for the part-related and process-related data (such as set-up files, part programs, component and process definitions etc.) and to provide sufficient capacity for any data archiving required for both the short term and the long term. These capacities are relatively straightforward to estimate, though it is certainly prudent to add a comfortable safety margin to whatever figure results, no matter how generous the calculation technique.

For many of these decisions, the FMS designer is to some extent likely to be in the hands of the computer system supplier. Parts of the specification will be derived by mutual consensus, while others will be based entirely on the estimates of the vendor's team. A significant degree of responsibility for the accuracy of these calculations must rest with the supplier, but with the best will in the world, estimating the performance of this type of system is not easy. If the software is not going to be developed on the hardware to be used for the FMS, then maybe some of these decisions can be left until more is known about the system, but the Project Manager will probably need to have some estimate of the hardware implications, if only for budgeting purposes.

Probably the best way to approach this problem is first, to delay it as long as possible, since more will then be known about the system, and second, to have as many people as possible calculate the requirements, preferably using entirely different philosophical approaches. This will add credibility to the figures eventually selected, especially if there is not too much variation in the overall conclusions. The only insurance policy one can reasonably have for this problem is to ensure that the hardware selected is easy and inexpensive to upgrade. With this knowledge, the task becomes somewhat less risky.

8.11 Concluding comments

Finally, it must be emphasised once again that the control software represents the most difficult and probably the single most expensive element of the FMS. It should be treated with great care and respect. The specification should be drawn up very carefully and then fixed. It should not be changed during coding unless absolutely necessary. A good measure of the excellence of the software specification is the number of times it needs to be changed — obviously, the fewer the better. Similarly, provision should be made at the outset for any planned extensions to the system — later unplanned additions could be extremely expensive. Certainly one way of minimising the impact of the software development is to plan for multiple applications. It is debatable how feasible such an activity really is, but if software costs can be amortised over, say, two installations instead of one with only a slight increase in cost, the advantages include easier maintenance as well as reduced cost. Additionally, the problem of sizing the computer system in terms of both capacity and performance becomes considerably easier and less risky during the second time implementation.

Adequate provision should also be made for the testing of the software. An attempt to put the FMS into production too early could lead to an unreliable FMS — that is, short-term benefits for long-term losses, not a good exchange.

9 Communications

9.1 Introduction

There are two fundamental differences between a flexible manufacturing system and a traditional automated system. The first of these is an FMS computer control or host system which facilitates the simultaneous near optimal utilisation of resources within the FMS. The second is the FMS communications system. However, it does not necessarily follow that an automated system which includes a communications network and a computer system is an FMS. The plant might simply be equipped with a sophisticated direct numerical control (DNC) system.

In many respects the differences between DNC and FMS are continuing to diminish as DNC systems become more and more sophisticated, a topic which will be covered in detail later in this chapter. However, DNC is frequently regarded as both a safe and sensible step towards FMS. The application of group technology to design and process standardisation perhaps offers the first step in FMS benefits, with DNC offering the second step by removing paper tapes from the shop-floor etc.

9.2 FMS communications requirements

An FMS usually contains a number of different communication systems. The reason why there are several separate systems within an FMS is because there are many different types of communications which need to be handled, all of which have significantly different characteristics. To date at least, it has been impossible to meet these broad requirements effectively and economically with one system. This situation might change with the take-up of General Motor's MAP communications protocol (a subject which is discussed more fully in chapter 12). If this is the case, which is by no means certain, it is unlikely to occur to any significant degree before the late 1980s.

Within all flexible manufacturing systems there are a number of 'centres of intelligence', between which communications of various types need to take place. Essentially, these are as shown in the following list:

- Operators.
- Workstation controllers (such as machine-tool, robot controllers, programmable logic controllers etc.).
- Workstation interface devices (that is, intelligent devices frequently used to interface a workstation controller to the FMS host computer system).
- The FMS host computer system, which could comprise several computers and their peripherals.
- Other flexible manufacturing systems.
- Remote factory computer systems, such as those responsible for plant-wide material requirements planning, computer aided design, inventory control etc.

While the operators are likely to communicate with all the devices comprising the FMS, figure 9.1 shows how the various devices would typically be interconnected.

In fact, there are eight types of communication which need to occur within a typical FMS installation:

(1) Workstation controller to operators.
(2) FMS host computer system to operators.
(3) Workstation interface unit to operators.
(4) Workstation controllers to workstation controllers.
(5) Between workstation controllers and their interface units.

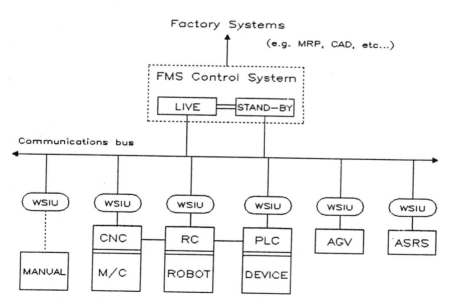

Figure 9.1   Typical FMS communications interconnections

(6) Messages between workstation interface units and the host computer system.

(7) 'Live'-to-'standby' computer communications.

(8) Host computer system to remote systems.

All these communications will tend to be bidirectional. For example, workstation controllers and their interface units will need to be able both to send and receive information from each other if successful communications are to occur. But not all these types of communications will necessarily be present within every FMS. Obviously, if there is no standby computer in the FMS, there will be no 'live'-to-'standby' communications.

9.3 Workstation-oriented communications

The communications which take place between the operators and various workstation control devices are obviously of a rather special kind. Their success is not only dependent on the quality and the design of the relevant electronics etc. but particularly on the user-interface within these devices. To some degree the importance of this topic was emphasised in chapter 8, all that will be added here is that particular attention should be paid to the user-interfaces on devices which are physically located on the shop-floor. This is because they are unlikely to be able to provide the same functionality and flexibility as computer terminals, and also because they will probably be operated in less than ideal conditions in terms of atmospheric cleanliness and ambient lighting conditions.

For example, workstation controllers and interface devices sometimes have only limited display capabilities, perhaps between one and eight lines of text. In these cases, careful thought must go into ensuring that the correct data is displayed in the most easy-to-understand manner, and that it is easy to obtain further information. Also, workstation controllers are equipped with displays which have been designed with a specific purpose in mind, namely the display of data pertinent to the operation of the workstation. The FMS and communication requirements are often an incremental, and probably secondary, burden in the mind of their designers, and hence it is quite likely that inadequate provision will have been made. Indeed, this is a further reason why so many FMS installations are equipped with workstation controller interface devices.

The communications between workstation controllers are usually of a rather special type. Frequently they are associated with the real-time synchronisation requirements of two workstations operating together. As such, they can be quite demanding. For example, a machine-tool and a robot, for the loading and unloading of parts or tools, need to be well co-ordinated if they are to function reliably as an integrated pair. Typically, this type of information is transferred by parallel interfaces with some form of optoisolation. The type of information usually being transferred will relate perhaps to the status of interlocks on closely coupled equipment. For example, the machine-tool might have to transfer the status of a fixturing device prior to the robot being able to carry out a task. A 'positive' signal from the device could be used to indicate that the fixture is open and hence ready to receive a new tool or component etc, and opposite sense signal might be used to indicate that the device has remained closed for some reason. Indeed, there could even be two such signals, effectively providing a check on each other — one saying the fixture is closed, the other saying it is not open, and so on.

Although some FMS installations have been implemented where these communications are passed through some intermediate device (such as a workstation interface device or perhaps a programmable logic controller), direct communications appear to be the most popular. Usually there is no processing associated with this information before it is comprehensible to the other workstation controller, hence there is little technical justification for the presence of an intermediate device. However, there could be some cabling and diagnostic advantages to be considered.

There are a number of ways in which this type of communication may be implemented. However, since typically the distances involved are relatively short, probably in the region of a few metres, and as most modern sophisticated workstation controllers incorporate, as a standard feature, the capability of generating and interpreting a number of such signals, this area of FMS communications is tending to become increasingly more straightforward.

Most of the communications which occur within an FMS are those associated with the transfer of information between the host computer system and the various workstation controllers within the FMS. There are many types of workstation controllers (for example, machine-tool controllers, robot controllers, programmable controllers etc.), and all can be purchased from a variety of different manufacturers. As they become more sophisticated, it will generally become easier to integrate these into a single FMS. At present there is frequently a need to provide an interface device between the workstation controller and the FMS communications network. Messages to different types of workstation controller then at least appear to be the same to the FMS controller, with 'translation' into the appropriate device-dependent code being handled locally by the workstation interface device.

In fact, interface devices may well be deemed necessary, regardless of whether they are actually needed to perform the task that their name implies. This is because they also provide the opportunity for an operator to interact with the host computer system via a standard user-interface (regardless of the location of the operator within the FMS and the manufacturer of the workstation controller) from the shop-floor. For this reason such devices are frequently used for both purposes, and maybe as purely an operator interface where there is need for operator interaction and yet no actual workstation controller, perhaps at a manual load/unload by or at a tool store, etc. One of the advantages of this approach is that maintenance can be simplified; for example, if all the workstation interface units (even those that do not have a workstation controller) are in hardware and firmware terms precisely the same, in the event of one failing, it needs simply to be unplugged and replaced with another. When the FMS control system detects the new workstation interface device, which could perhaps sense the address from some appropriately connected terminals within the input socket to the FMS communications system, the required information to facilitate communications with the relevant workstation controller would be downloaded automatically. From that point onward, the FMS control system would not need to concern itself with the precise type of the workstation controller. However, this situation is only likely to exist if all the workstation interface units are the same, since then it is justifiable for the user to have a spare unit always available.

Communications between the FMS computer control system and the workstation controllers, and possibly their interface devices, take place to provide a number of facilities, essentially as follows:

(1) To transmit part programs, parameter files (such as tool offset tables etc.) and possibly other text files, in both directions.
(2) To permit transfer of status information from the workstation controller to the FMS computer.
(3) To enable the FMS computer to issue commands to the workstation controllers.
(4) To facilitate operator interaction with the FMS computer from the vicinity of the workstation.

Transfer of part programs and associated data is a basic function of all factory communication systems. The main difference between FMS and DNC communication systems is essentially that the ability exists within the former to exert a considerable amount of control over the workstation controller. By definition, this is a facility which is not required of a traditional DNC system, whose main purpose is to eliminate part programs on consumable media, such as paper tapes, and improve efficiency. With regard to the transfer of status information, many of the more sophisticated DNC systems do share this capability with FMS communication systems.

A substantial proportion of the burden on the communications system is likely to be that associated with the transfer of a number of types of files typically between the host computer system and the workstation controllers. These might be for the machine tools, the robots, or perhaps programmable logic controllers, etc. It is highly unlikely that the programs will be of the same format — some will comprise alphanumeric characters while others could consist only of binary characters. It is likely, therefore, that any one

FMS communications system will have to support the transmission of data in a variety of forms. Usually it will be necessary to send some associated data with the part programs. For a machine-tool this could perhaps be a tool offset table, for a programmable controller this could be a recipe file, while for a robot it could be a parameter file (numbers which will be used by the robot part program to perhaps show where components are located within the operating area, how many parts are on an incoming pallet etc.).

In addition, it will probably prove desirable to be able to send text files. These might be files which could not be interpreted by the workstation controller, but could perhaps be used by the operators when setting up the workstation to manufacture a different type of part to that previously produced. Such a file would typically contain information such as which chuck jaws should be used in a lathe, and where they might be found, settings of various ancillary equipment such as robot gripper operating restrictions, selection of sensor banks etc. Graphics display data may also need to be sent. Within more sophisticated environments, all these support files could be made 'interactive'. If this is the case, such files are often used to form the basis of the way in which recovery of equipment following failure is achieved.

Though there is certainly a need for the FMS host computer system to be able to send all these types of files automatically (that is, without any operator intervention), assuming the correct equipment status exists to permit such transfers, there is also the need to instigate the sending and receipt of these files in an essentially manual mode. In this case the transfer might be initiated from either the workstation controller, its interface device or the host computer system. Such transfers should, like their automatic counterparts, only be permitted to occur if both workstation and host computer system status is appropriate, the identification of this being one of the tasks of the host computer.

The need for such manual transfer will vary in importance between different systems. For example, in an FMS where part programs are frequently modified on the workstation controllers, it will be desirable to send these back to the host computer system for archiving, and for use when the part is next produced. Similarly, during recovery of the FMS after an error has occurred it will probably be necessary for the operator to instruct the host computer system as to the new status of the failed part of the FMS. This will enable the host computer system to create and send a new parameter file etc. to the relevant workstation. Also, after perhaps a tool has been changed or a component inspected, it is essential that new data is sent to the host computer system.

The success of a flexible manufacturing system will usually depend on its ability to collect and react in a timely manner to data relating to the real-time status of the system and to make adjustments when events do not occur quite as was expected. Depending on the equipment and processes within the FMS, the status information that is likely to be important will vary. Typically, this status information would include:

- Workstation powered up
- Auto communications enabled
- Manual communications enabled
- Workstation in manual, automatic or edit mode
- Workstation busy
- Workstation in single block mode
- Feed rate override on
- Urgent alarm (possibly with type)
- Non-urgent problem (possibly with type)
- Waiting for another workstation or operator
- Optional stops enabled
- Part count
- Active part program name and block number
- Workstation attention light on/off
- Operator requests that a recovery be instigated

Similarly, the type of commands that the FMS control computer might well wish to pass on to the workstation would be as follows:

- Change workstation status to DNC off
- Place workstation in auto mode, manual or single block mode
- Enable optional stops
- Switch workstation attention light on or off
- Display the following message
- Execute a pause until further notice
- Execute an emergency stop (not desirable in many systems since this action might create a hazard for operators)
- Clear part program memory
- Receive the following program
- Send the following program
- Receive the following parameter file
- Send the following parameter file
- Enable or disable the following program
- Receive the following text file
- Display the following text file

Examples of the communications-related interactions which an operator might typically wish to have with the workstation controller, or the workstation interface device, are as follows:

- Display the set-up file
- Cancel previous request

- Terminate previous request
- Send an operator message
- Start a recovery
- Note a tool change
- Change workstation operating mode
- Prepare workstation for start up or close down

All these requirements are interrelated, and to some degree application-dependent. They are all a reflection of the status changes that the FMS control computer is likely to have to monitor and react to. However, putting all these messages and interactions together in a form in which they will reliably perform a particular function is quite complex and requires the input of both skill and experience if success is to be guaranteed. For example, if the host computer system wished to send a file of some description to a workstation controller or its interface device, a typical (but by no means the only) train of events might be as follows:

- Computer checks workstation controller (and/or interface device) is in the correct state to receive messages.
- Computer sends messages to ensure that the workstation controller is likely to remain in the correct state to receive communications.
- Computer sends message to delete any files currently stored.
- Computer sends message 'prepare to receive the following file named PARTNAME.OPN'.
- Computer sends first file block.
- Computer sends next block (repeated as needed).
- Computer sends last block.
- Computer sends 'file transfer complete message'.

The above represents the messages which might be sent from the host computer system. All these messages would have to be appropriately acknowledged by the workstation controller and/or its interface device. Usually this acknowledgement is sent in the form of the 'ACK' character, which would be used to denote that not only had the previous message been received and understood, but also that the device was ready to receive the next message. Alternatively, if a 'NAK' was sent, this might mean that a message had been received, but that it was not understood (it might have been corrupted during transmission). The 'NAK' would mean that the receiving device was ready to accept another message, probably a repeat of the one which was not understood.

A number of safeguards are usually added to this process of message acknowledgement, for example, there might be a time-out on the delay between the control computer sending a message and the device replying, either with an 'ACK' or a 'NAK'. If the delay exceeded a pre-set interval, the computer would perhaps issue a message to the FMS operators that transmissions to that particular device had ceased to be successful.

Similarly, the number of times that the control computer will repeat the transmission of a message for the benefit of a device which did not receive the message correctly would normally be restricted, perhaps to, say, four or five. In the event of a message being re-sent this number of times and still not being received correctly by the device, an error message would be issued to the operators. All of these features would need to be provided as an integral part of the FMS communications system protocol. A number of well-proven proprietary protocols and supporting hardware/software are available. Suffice to say, it is usual to monitor these transmission errors and compile them as part of the FMS statistics. This enables the System Manager to monitor the reliability of the FMS communications system.

To simplify the understanding, development and operation of this type of communication system, it is usual for all the communicating devices to have a clearly defined 'master–slave' relationship. In some systems, this master–slave relationship is dynamic, varying according to the needs of the system at that time. While in others it is fixed during the design of the system, and remains so throughout the operating life of the FMS.

Essentially, the master–slave relationship between two or more communicating devices dictates which device is able to initiate an interaction, and which has to respond to the requests of the other. For example, if, as is frequently the case, the host computer system is the designated master, it must instigate communications. If the computer wants to, say, send a file, it simply instructs the workstation controller to prepare to receive the program, assuming that the status is correct. However, if the workstation controller wants to send a file to the host computer, perhaps for back-up purposes, it is rather more difficult to initiate this process. This is usually achieved by allowing the workstation controller and/or its interface device to place a message in its output buffer such that when the master next asks for the status message of the controller, the fact that a request is outstanding is noted. The FMS control computer would then be able to respond by asking the controller for details of the outstanding request. One of the problems with this is that if there is a delay before a status message is requested, there is nothing the workstation is able to do to speed up the transmission process. An operator, therefore, standing by the workstation waiting for the transmission to take place, might become irritated.

This raises again the issue of response time within the FMS, and in particular the rate at which workstations are polled — that is, interrogated by the FMS control computer system. This is a factor which substantially influences the selection of the host computer system for the FMS. Once a status message requesting further communications activity has been collected by the host computer, there is no reason, unless the computer is busy with another computer-intensive task, why the remainder of the transmission cannot become a fairly high priority task, and as such be completed appropriately quickly. However, ensuring that this potential delay before transmission start-up is acceptable, could be a costly process.

Although the master–slave relationship described above is very popular within flexible manufacturing systems, it is by no means the only way in which such industrial communications systems can be made to operate. Indeed, the approach which is most appropriate will depend on the type of communication system being used and the host computer within the FMS. Also, as mentioned previously, the master–slave relationship could alter dynamically to reflect the requirements of the system at any particular time. This is effectively what occurs with 'token passing' networks, where a logical token, or right to communicate, is passed in turn from device to device according to the rules of a well-defined control algorithm.

9.4 Some implications of standby computers

Where resilience to computer failure is required within an FMS, perhaps to facilitate continuous 24 hour a day operation, this is usually achieved by providing a duplicate host computer system, a concept which was introduced in the previous chapter. The communications between the live (running the FMS) and standby (the back-up) computers are responsible for ensuring that, at any time, the standby computer would be able to take over control of the FMS (as shown in figure 8.4). These systems are currently installed in one of three ways: either as a 'cold' standby, a 'warm' standby or, finally, a full 'hot' standby system. Though this final type of system used to be achieved with specially developed software, an alternative route is in the process of being developed and although not yet applied within any FMS environments, does show promise of being applicable. This is 'cluster' technology, where a group of computers are collectively responsible for the operation of a system. When one unit fails, the remainder continue to operate the complete system, possibly in a slightly degraded mode. Although the current cost of the technology makes the approach expensive, this situation is likely to change quite rapidly during the next few years.

To implement a 'cold' standby computer control system for an FMS is the least complex way in which a back-up facility can be provided. Once the live computer has failed, it is essentially the operators' responsibility to start the standby, or secondary, computer and restart the FMS. This may be a relatively straightforward task if the FMS status files available to the standby computer precisely reflect the current status of the FMS (as might well be the case if the previous live computer's disk packs have to be physically transferred to the back-up computer system). If, as is likely, the status files are out of synchronisation, some manual intervention will be necessary to ensure that the back-up computer is eventually able to continue running the

FMS. The operators and the computer may well have to go through a question-and-answer session to bring the database up-to-date. This would need to be started at the termination of the last event completed by the host computer which failed, and continued through to the last event which actually occurred within the FMS. This would eventually result in status files which reflect the new status of the FMS. It is likely that throughout this process the FMS will need to be kept in an inactive mode, in order that the FMS is maintained in a known state. The magnitude and complexity of this recovery task is likely to depend on how quickly the status files become out-of-date, and how difficult system recovery is likely to be in physical terms. Suffice to say, it could represent a significant problem, hence the interest in the other forms of standby systems.

The communications which need to take place between two computer systems which, collectively, are responsible for the control of the FMS on a hot or warm standby basis, are quite difficult to implement. The inter-processor link between the computers is usually relatively simple to implement, in physical terms, since the computers are probably of the same type, and also because they are likely to be located in close proximity to each other. However, the software co-ordination of the two computers across this link might not be quite so straightforward to engineer. Since the warm standby system may essentially be regarded as a subset of the rather more complex hot standby, most of this discussion will be devoted to the requirements of a hot standby system, the preferred architecture in most modern systems designed to operate for 24 hours each day. Though with evolving computer technology providing more reliable computers (resulting in fewer failures and less maintenance time being required), this situation may change, substantially reducing software costs and risks.

There are essentially three elements to the provision of either a hot or warm standby facility. The first of these is the 'watchdog' facility (this may not be present in warm standby systems). This is hardware and/or software which enables each computer to check whether the other is operational or not. At a certain time interval, one computer will send a message to the other, and if the correct reply is received within a pre-set time, it is concluded that the other computer is active. If there is no reply, a computer has failed; in which case, if appropriate, the responsibility for the control of the FMS will be changed — that is, the standby computer will become live.

Conceivably, the inter-processor link connecting the two computers could also fail. If this is likely to be a frequent occurrence, another means of checking the viability of the other computer must be provided, or the situation could arise where both computers are simultaneously attempting to control the same FMS. A solution would be to use a 'bus' or 'ring' communications network. These approaches will be discussed later in the chapter.

While the two computers are active, generally two types of data will need to be transferred across the inter-processor link. (This may or may not be the same link as used for the watchdog timer.) This is non-time-critical data, such as machine-tool part programs etc., and time-critical-data, such as the real-time status of the FMS. Transfer of non-time-critical data which is being archived is not in itself particularly difficult, the key problem to be addressed being to ensure that data integrity is maintained — that is, an old version of a file is not allowed to update a newer version (a problem particularly relevant in dual computer control systems). The on-line transfer of time-critical data such as the FMS status is more complex.

The problem with status transfer is essentially one of ensuring that when the back-up computer takes over control of the FMS, it actually has all the current data available to restart and continue running the FMS. The shorter the mean time between status changes within the FMS which need to be monitored by the host computer, the more difficult it will be to provide complete hot standby. One method frequently used to overcome this possibility of losing status messages is to have all the messages sent to both computers. This imposes certain constraints on how the communications architecture is designed, and how individual messages are treated.

Usually, actions will be subdivided into both requests for action and the commands which are issued in response to these requests. Although both computers would be able to receive the requests, only the live computer would be permitted to issue the response. Typically, a master–slave relationship exists between both the computers, similar to that between the host computer and the other devices within the FMS. However, this need not prevent the standby computer from monitoring all the messages received, and sent, from the live computer.

Systems such as these enable the standby computer to keep a close watch on the progress of the live computer. In the event of the live computer failing, this probably being indicated by the failure of the watchdog timer, the standby would know which devices had outstanding requests for action, and therefore immediate takeover should be possible in most cases. However, with a shorter time interval in which the response must be issued by the live computer, and especially if this interval is of the same order of magnitude as the operation of the 'watchdog' timer, then the risk that the standby will be unable to takeover the responsibility of the live soon enough to issue the response is substantially increased. In such an environment, and it is not too difficult to create circumstances such as these, a response may not be issued and the real-time status of the FMS may be lost in certain areas. In this situation, there is no alternative but to request operator intervention to confirm the status in the appropriate sectors of the system. Hopefully this intervention will be at a minimal level and the new live computer will be able to continue normal operation of the remainder (that is, most) of the FMS.

Suffice to say that this difficulty of not being able to fully guarantee the immediate control takeover by the FMS standby computer, especially when considered with the substantial incremental costs and risks associated with developing the software needed to operate these systems, is one of the reasons that hot standby systems are being implemented less frequently.

9.5 Communications with remote computers

The communications which might take place between an FMS control system and another computer system could be quite varied in nature, and possibly quite difficult to implement. As computers have generally become more powerful, and communications more standardised, the difficulty that used to exist in simply providing for data transfer between two or more computers, say of different manufacture, has largely been eliminated. This means that to pass what are essentially passive files is not particularly difficult. For example, if part programs for machine-tools within the FMS are being prepared on a remote CAD system, it is a relatively straightforward matter to pass these to the FMS control system. Usually, they would automatically be sent to the live computer, and then perhaps archived on the standby.

Such static production support files could also be sent from the FMS control system to a host computer, perhaps for back-up purposes in the event of there not being any other data back-up facility such as production statistics, support files etc. Ensuring the integrity of data transmitted is the real technical issue involved. Though having received the data correctly, the respective computer has to be provided with appropriate software to know what to do with it.

Where problems can arise is when it is necessary to transmit active, say, FMS status or work schedule files. When being sent from the host computer system to a remote computer system, all that is needed is an accurate copy of the data, which of course will almost immediately be out of date, since the status of the FMS will have continued to change even during the transmission of the file. However, assuming the copy or 'snap-shot' can be made without affecting the operation of the FMS, it can then be transmitted to the appropriate computer.

However, if such a file is being sent to the FMS control computer from a factory remote system, the situation is slightly different. Although the transmission of the file might well be relatively simple, problems can arise when attempts are made to utilise the data within the files. This is essentially because a certain FMS status may have been used as the basis on which the work schedule was prepared. However, the status of the FMS is likely to have changed, making the schedule unworkable. This presents an interesting and potentially complex problem which has far-reaching implications for the ways in which it might be appropriate to structure manufacturing decision-making responsibility within a CIM environment. The most frequent conclusion is that the simplest way in which this situation may be resolved in a generally applicable manner is to distribute the responsibility for internal scheduling of the FMS concerned to the relevant host computer system. This approach not only has the advantage of reducing the size of the scheduling problem, thus facilitating the identification of an acceptable work schedule, but also provides the means for minor system perturbations to be absorbed locally, without the need to concern the remote system, which may not be operated on a continuous basis. Such an FMS work scheduling system would only need to receive guidelines concerning the orders required. For example, how many parts are required at a particular time, the earliest acceptable start date and the latest acceptable finish date. The decision as to when the job was actually run would be left to the FMS computers. Another advantage of this approach is that it caters for both the environments where there is not a remote MRP computer.

9.6 FMS communications architectures

There are essentially four types of overall communications architectures which are commonly applied within FMS environments: the 'star' network, the 'ring', the 'bus' and the 'tree'. All have their advantages and disadvantages in certain environments, though the star network is probably the older of the approaches, since the technology required for the other techniques has not been economically and commercially available until recently.

A summary of the architectural principles involved in these approaches is shown in figure 9.2. Figure 9.3 presents an indication of the likely cost and performance implications of the various approaches to communications.

It is important to note that the cost of installing these systems should not be underestimated. Certainly the fewer cables that are involved the better, since, for example, fewer cable ducts would be needed, thus reducing the amount of work needed to lay the cable both from the point of view of physical installation and of separating power lines from communication lines. However, the number of cables and their ease of installation are not the only factors which need to be considered. Also, there are issues of line balancing, the number and location of repeaters etc. More strategically, however, when selecting a communications system for an FMS, it is essential that the implications of factors such as the availability of long-term support, extendability, market and technological trends and reliability are also taken into account. Once a system such as this has been installed, it might remain with the company for a long time.

The tree network is probably the newest addition to FMS communications architectures. It is a derivative of the bus structure and is that selected to be used in MAP environments (see chapter 12). At present, a sufficient number of such systems have not been installed to permit many comments to be made. The technology is new enough to be prohibitively expensive, but in a few years'

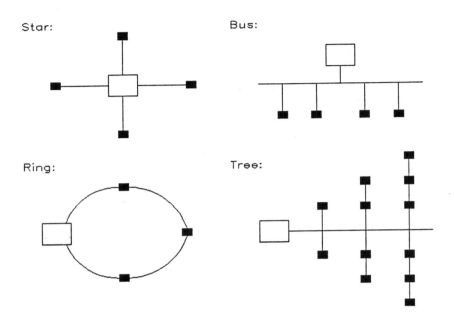

Figure 9.2   The main types of FMS communications architecture

TYPE	INSTALLATION			INTERFACING	
	Length	Cables	Cost	Speed	Cost
Star	Long	Many	High	Slow	Low
Ring	Short	Few	Medium	Fast	Medium
Bus	Short	Few	Low	Medium	High
Tree	Medium	Medium	High	High	High

Figure 9.3   A comparison of communications architectures

time, it is likely that this type of architecture will become less expensive to implement, and hence commonplace.

Probably the single most important factor of these is that of reliability. After all, once an FMS has been installed, the key requirement is to keep it operational for as large a proportion of the working time as is possible. The inherent reliability of star networks is extremely high. This is largely because each device is connected to the control computer by its own communications line. In the event of there being a failure on this line, only one device is affected. If there is a board failure at the control computer, typically only eight devices would be affected.

The mere fact that each individual device is connected directly to the host computer system (or perhaps to a switchover unit if a dual computer system is being used) does present a substantial cabling task. If a switchover unit is installed within a hot standby environment, the task is even greater and will have to be capable of being activated by the watchdog timer — an added complexity for hot standby systems.

Probably the bus network is the next most reliable since, typically, it will have between ten and sixteen devices connected to the one line, all being addressed by the same board within the control computer.

The ring network is to some degree likely to be the least reliable configuration since all the devices within the FMS are connected to the same (possibly co-axial) cable. If the cable was broken, communications within the entire FMS would be lost. If a multi-segment approach with repeaters has been adopted, then the repeaters can be configured in such a way that many of the segments will remain operational after a failure, though this may not be enough to keep the FMS operational for more than a short period of time.

The issue of long-term support is somewhat variable, depending on the particular supplier selected to supply the network since there are a substantial number of variations on the three themes mentioned above. All are supported in some form by various suppliers. Probably the bus approach is, to date, the least popular. Although the star network has been the traditional approach for many years, the ring architecture is probably the approach which has received most publicity recently because of the many quite successful applications within the office environment. Whether these successes will be repeated on the shop-floor remains to be seen.

Extendability really comprises two issues: first, the ease with which a new device may be added to the network and, second, the ease with which an existing device may be reconnected following a physical relocation.

With a star network, the cost of integrating another device at the control computer is relatively low if another board does not need to be purchased. Then it is purely a matter of laying the cable to the device. With the bus and ring networks the problem is slightly more complex because existing cable will need to be re-routed and possibly more repeaters added.

Concerning the relocation of an existing device, the above comments relating to the bus and the ring networks essentially apply again. However, where the star network is concerned, it frequently proves easier to lay a new cable rather than attempt to use a section of the existing cable. Obviously, the impact of these factors is going to vary according to the likelihood of the FMS being extended or altered. If neither of these is likely to occur in the forseeable future, following installation of the system, the relevance of these considerations will be small.

There are essentially two media which are used to transmit the various data. These are traditional wire or cable and optical fibre. As might be expected, there are advantages and disadvantages associated with the use of both of these media. Wire tends to be the more traditional and, to date, more popular approach. However, advances are continually being made in communications technology, and while the cost of using optical fibres is continuing to decrease so its reliability and flexibility continues to increase. Certainly, optical fibre does have the considerable advantage of not being prone to interference from devices (such as arc welders etc.) which may be located close to the FMS environment. However, optical fibre does also tend to be rather more fragile, and not quite so easy to install since, for example, it can neither be joined so easily nor curved in such tight radii as wire. For the near future, at least, wire (probably in the form of twisted pair of cables or coaxial cables) looks likely to remain the most popular medium for FMS communications.

Indeed, there are really only two significant disadvantages associated with the use of wire for FMS communications. First, the bandwidth or capacity of the cable is somewhat lower than that of optical fibre, and messages are prone to corruption from interference. Indeed, the capacity limitation is probably the single most important factor which has led to the enthusiastic development of optical fibre. Certainly, in FMS environments where the communications traffic is high, such a system might well have operational advantages which outweigh the incremental expenditure necessary.

Nevertheless, as intimated earlier, if wires are being used as the means of supporting the FMS communications, it is essential that messages sent are checked on receipt for data integrity. This can be done in a variety of ways. The most frequently adopted approach is to incorporate a combination of two techniques.

First, a parity check can be made on the message. This is a relatively simple procedure which checks that a feasible set of characters has been received, essentially by checking whether the character count is odd or even as defined within the system. However, this will only detect whether a message is feasible or not. For example, if two characters were lost, the parity count might still be odd or even as required. In other words, the message might still be understandable, but nevertheless incorrect.

Second, to help overcome the above problems, a special algorithm is used which multiplies the values of the constituent elements of the message and appends what is called a checksum to the message. When the message is received, the algorithm is again used to multiply the values of the characters comprising the message, and the resulting checksum is compared with the one contained within the message. If the two values are different, the message is in error, a 'NAK' would be the device's response and, if the retry count was still acceptably low, the message would be repeated. The processing overhead involved in this type of procedure is relatively small, while the level of certainty that the correct message has been received is increased significantly.

Regardless of the architecture and functionality of the communications system ultimately selected, it is important that adequate provision is made for the training of both installation and maintenance personnel. It is imperative that as the communications network is installed it is tested for integrity in terms of both the quality of the installation workmanship and the medium being used. Also, the level of interference being received must be checked together with the location/operation of repeaters etc., to ensure that signal strengths are adequate and well balanced throughout the system. The latter is particularly important if the communications system is likely to be expanded at a later date.

9.7 Concluding comments

When the FMS is being commissioned, it is important that the level of confidence in the communications system is high. It is likely that fault diagnosis within the FMS will be difficult anyway: not being sure where the fault is likely to have originated is purely an unnecessary confusion. The key to successful commissioning of any complex system is to ensure that only one element of the system is being tested at any one time and certainly correct operation of the communications system is a fundamental component of this procedure.

When selecting an FMS communications system, it is important to bear in mind not only the current requirements of the system in hand, but also the likely future requirements of both this FMS and the rest of the company. In a time when, for example, the GM MAP protocol is rapidly gaining momentum (though at the time this book was being written not much MAP-compatible hardware was available, and that which was tended to be prohibitively expensive), it is important that the relevant influences are considered. While it might not be practical to consider incorporating MAP at the present time, it would be foolish to preclude the possibility of its use later. This means that due consideration should be given to the possibility of providing a migration path to MAP (or whatever, if any, communication systems does establish itself as a standard).

10 Financial Justification

10.1 Introduction

This chapter is written for the benefit of project managers and engineers who might have the responsibility for the design and implementation of an FMS. It is not intended to provide all the necessary theory on how a full financial appraisal should be carried out, but more to ensure that if such an analysis is required, and is approached as suggested, it will prove to be both comprehensive and acceptable.

10.2 Some background

Interestingly, the discussions of how a flexible manufacturing system should be financially justified appear to be almost more popular than FMS technology itself. This is probably because the FMS 'band-wagon', upon which many people jumped in their enthusiasm to take maximum advantage of government grant schemes offered in many countries, has been seen to present not too 'comfortable a ride' if the correct precautions are not taken, both technically and financially. Indeed, it has been suggested that many of the early FMS installations are not financially viable. This may or may not be the case; it is not always easy to calculate the complete impact accurately. However, if a system is deemed not to be financially viable, this may not be the fault of the technology employed so much as the overall system design and the financial analysis.

One of the many reasons why it is difficult to justify an FMS financially (both before and after installation) is because people still are determined to look upon such a system purely as a replacement for a selection of traditional manufacturing technology. Just as an FMS is many times more complex than a CNC machine-tool, so trying to justify an FMS is considerably more involved than justifying a straightforward plant replacement. Indeed, continuing with this analogy, the typical CNC machine-tool represents a small component within a larger system, while an FMS is a complete manufacturing system in itself, which is operating on one set of rules, within an even larger production system, operating with another set of rules.

When a decision has been made to invest in FMS, it is usually a long-term strategic decision, not a tactical adjustment within a strategy. Unfortunately, much of the impact of this decision is frequently not considered in sufficient detail. Many of the considerable advantages which result from well-conceived FMS installations are extremely difficult to quantify. This is true of virtually all strategic investments, whether it be a change in computing policy, a new chairman of the board, or purchasing an FMS. This problem is further compounded by the fact that traditional financial analysis techniques, as used by accountants worldwide, are not particularly appropriate to the appraisal of this type of highly automated system, since they do not measure all the variable outputs which can result.

10.3 Some strategic issues

When first embarking on the FMS route, and before becoming submerged in numerous financial analyses, one of the most important considerations is to quantify the impact of the substantial investment on the corporate goals of the company. For example:

- Where will the company be in 5 or 10 years time?
- Is this investment going to provide an advantage over the company's competitors, and if so how?
- Will it capture a greater proportion of the market for the company's products?
- Will it provide the ability to respond more quickly to market changes?
- Will quality improvements and consistent output provide competitive advantages?
- Will the publicity gained from the company using advanced technologies result in incremental sales?

All these issues are of fundamental importance to the justification of any FMS. As is apparent, the list of relevant issues is already long, and no mention has yet been made of the justification of the selected technology. Many of these issues are of such a type that they are difficult to quantify accurately. Possibly, an attempt at quantifying them might even be misleading. An important consideration is that of what actions the competitors are taking, and why these are felt to be justified. If the competitors are in the same country, have similar cost and benefit structures, and are seen to be investing in FMS, is there any other way of staying in business, other than to follow suit with a more efficient system?

Of course, not all the implications of FMS are necessarily advantageous. For example, such an investment could be viewed (usually incorrectly) as a means of replacing labour. This could give rise to industrial relations problems. Certainly in some countries, the lack of creative investment is often, quite correctly, blamed for the demise of many companies, particularly so within the ailing machine-tool industry, where the difference in manufacturing techniques between the West and Japan is substantial.

There is no doubt that any FMS must be an integrated part of a clearly defined long-term manufacturing strategy aimed at enabling the company to meet its corporate goals and ambitions. Unfortunately, both the goals and the means of obtaining them are likely to be moving targets. Nevertheless, these strategies will at least enable the FMS designer to establish some guidelines as to where and to what extent, for example, the system needs to be truly flexible. Perhaps more importantly, it will help to ensure that the FMS will easily become an integral part of the current and planned manufacturing capability of the company, hence putting the project in perspective, both financially and technologically.

It is often the case that FMS Project Teams are too enthusiastic about becoming embroiled in the glamorous aspects of FMS technology — parts selection, design of the required hardware and software, and identifying new technology which could be crammed into the system etc. This is very much to the detriment of the long-term interests of the FMS itself, especially within a background of traditional capital investment appraisal techniques. It typically leads to the main advantages that could be obtained from the investment either being ignored or lost. Usually the FMS Project Team is too far removed from the overall strategic business issues which are being addressed by upper management. This is an important problem which should be addressed at an early stage by the Project Manager. If the investment is mis-directed, while eventually an FMS might be seen as a technical success, in overall market terms the result may be to provide the company with nothing other than relatively superficial operational advantages, which probably could have been obtained in a more appropriate manner.

Certainly, by simply applying the cost/benefit formulae that have been used so widely in the past, it is quite easy to miss what could be the real justification for the whole project. An especially dangerous situation exists if the appraisal technique is being used to compare a number of different investment alternatives. Without doubt, the analysis techniques frequently used tend to be myopic and limited to 'bottom-line' implications. What is needed now are techniques which provide financial analysis on a total systems basis. After all, this is the way in which automation is now applied. Otherwise one is left with an appraisal of operational advantages and disadvantages of one system compared with maybe an existing system running under entirely different rules. The analysis may at best expose only a small percentage of the relevant facts.

Nevertheless, while strategic advantages are of immense importance, this does not mean that a detailed financial analysis should not be carried out. Just as an FMS tends to expose problems in other areas of a production system, so wider-ranging financial analyses can reveal problems elsewhere within the business structure. For example, it might be concluded that investment in an FMS apparently results in an increase in the cost of manufacturing parts. Given the strategic advantages that could be directly or indirectly attributed to the investment, this could be deemed to be an acceptable situation. But prior to this, one of the questions which should be asked is whether the accepted cost of parts actually reflects their true production cost to the company. When an enquiry arises for the development of an FMS, the first step which should be taken is a total review of the company concerned, in particular its strategies and its currently installed facilities. Only when this has been carried out will it be possible to initiate a financial appraisal of an FMS with a reasonable degree of certainty as to the validity and applicability of the results of the analysis.

10.4 Performing the analysis

In theory at least, carrying out a financial analysis should not be particularly difficult, although this often proves not to be the case, usually because the base data is not available or else cannot be guaranteed to be accurate. Certainly, it is not sensible just to perform one analysis, manually. Instead a number of analyses using different approaches should be carried out, preferably using computer techniques wherever possible. This will not only remove much of the drudgery associated with calculating say 'Net Present Values', but more importantly, it will allow the analyst to perform a detailed sensitivity analysis with different investment possibilities. These systems, many of which are available to run on microcomputers, whose use has been strongly advocated throughout this text, enable both quick and comprehensive comparisons to be made of alternative investment possibilities. Typically, these might comprise: 'do nothing', 'replace the existing facility like-for-like', 'buy CNC' and 'buy FMS'.

Such investment appraisal systems enable the analyst to study carefully the financial implications of the various equipment additions necessary to progress from one alternative to another. It may well transpire that a carefully planned progression from one level of automation complexity to another will not only ease the introduction of the new technology but also allow some of the benefits realisable from the early phases to assist with the funding of the later stages.

As mentioned previously, it is not the intention within this chapter to pontificate on how to perform the necessary financial analyses. Probably most readers will either already be familiar with these time-honoured techniques or have access to some of the many publications devoted entirely to the topic. Nevertheless, a brief FMS-oriented overview of the major techniques will be included, accompanied by some suggestions for the presentation of results.

There are a considerable number of appraisal techniques used for FMS evaluations. However, these are essentially derived from some five basic approaches, variants then being produced in an effort to improve the applicability of the selected formulae. These approaches are as follows:

(1) *The pay back method* — the time needed to recover the investment from the incremental profits generated.
(2) *The return on investment method* — the percentage of the investment represented by the profit generated in a 'normal' year of operation.
(3) *The net present value method* — which takes into account the cashflows generated throughout the whole life of the project.
(4) *The internal rate of return method* — which is similar to (3) but is used to find the percentage interest rate which will result in a zero net present value.
(5) *The MAPI method* — which is used primarily for replacement investments, is also based on the net present value approach, adjusted for capital utilisation.

To pursue an example of how such an analysis might be carried out, it will be assumed that a discounted cashflow (DCF) technique is to be used. Despite the proven difficulties with the use of this analysis technique, it remains popular, as demonstrated by the fact that it is still included in all accountancy examinations and most good engineering courses. Whether the former is the cause or the result of the latter is a debatable point. Either way, this does not alleviate the problem that these techniques and the meaning of terms such as Net Present Value (NPV) and Internal Rate of Return (IRR) are learned by engineers 'parrot fashion', and often with little depth of understanding of their real implications and weaknesses. In fact, analytically the two concepts are relatively simplistic. Both essentially rely on the principle that the purchasing power of money will inevitably reduce as time passes.

In Net Present Value calculations, an annual percentage decrease in the value of money is supplied as an input by the company. This 'minimum acceptable rate of return' is usually used to represent the percentage increase in the value of money which the company would normally expect to obtain. If the net inflows resulting from the investment, when discounted over time, are greater than the net outflows, the project is deemed to be profitable. When projects are being compared, a 'profitability index' is sometimes calculated. This is a comparison between the total present value of the cashflows with the amount of the original investment, namely a measure of the percentage increase in the investment that the present values of the cash-inflows represent:

$$\text{Profitability index} = \frac{\text{Present value of cashflows}}{\text{Initial investment}}$$

As mentioned previously, the Internal Rate of Return method is in essence similar to that of the Net Present Value. However, instead of a minimum acceptable rate of return being defined as an input, a rate of return is calculated which balances the cash-inflows over time with the cash-outflows over time. This tends to be an iterative process, and quite laborious if carried out manually. Needless to say, if the resultant internal rate of return is greater than the company's minimum acceptable rate of return, the project is at least deemed to be profitable.

Regardless of which technique is employed, one of the most serious problems likely to be faced by the Project Team in preparing for the analysis is that of obtaining the base data for the generation of the annual cash-inflows and cash-outflows. Needless to say, great care should be taken to ensure that this data is as accurate as possible, and where doubt appears this should not only be clearly identified, but also a sensitivity analysis should be performed to assess the relative importance of the uncertainty. Ultimately, both techniques have their drawbacks, and ideally as many alternative approaches as possible should be used when considering a significant investment. This reduces the probability that the biases of the appraisal techniques are permitted to bias the investment decision.

Since it is generally accepted that the NPV technique is likely to prove superior in most common practical situations, it is this which will be used in the following relatively straightforward examples. In its simplest form (not including allowances for inflation and taxation) the Net Present Value of a project may be defined as follows:

$$\text{Net Present Value} = \frac{\text{Future net inflows}}{(1 + i)^n} - C$$

where

i = annual discount rate
C = initial investment
n = project duration (years)

Unfortunately, this rather simplistic representation is considerably confused by the fact that the majority of the values are not known with any degree of certainty, especially at the earlier stages of the project, which is when most such analyses are carried out. For example, it is an accepted fact that the costing of software developments is extremely difficult. Since the development of the control software for an FMS represents a significant proportion of the total cost of the system, errors in calculating this amount will have a substantial influence on the validity of the entire analysis. Indeed, all the terms are subject to uncertainty in their actual values, or alternatively, are affected by fluctuations of unknown magnitude (such as inflation rates, rates of corporation tax etc.), an unfortunate state of affairs which does not help analyses such as these.

Year	Cashflow	20% PV factor	Present value
0	(3,000,000)	1.000	(3,000,000)
1	0	0.864	0
2	650,000	0.694	451,000
3	1,100,000	0.579	636,900
4	1,500,000	0.482	723,000
5	2,000,000	0.402	804.000

Total cash outflow		(3,000,000)
Discounted total cash inflow		2,615,000

Project Net Present Value	(385,000)

Figure 10.1 Typical net present value calculations

Nevertheless, for an FMS which is estimated to cost 3 000 000 (all amounts being quoted in unitless format, with '(−)' denoting net cash-outflows), the principle of the NPV analysis is essentially as shown in figure 10.1.

The 3 000 000 sum would typically represent the best estimate for that time frame of the total capital needed to implement the system. If there were some grants available for this type of project, a proportion of the funds might be recoverable, though some incremental costs might be incurred in their place (for example, public demonstration of the system etc.).

Suffice to say, a five year analysis period has been used to minimise the length of the investment cycle and hence the analysis; but it is also not unrepresentative of an appropriate investment horizon for an FMS.

This superficial analysis shows that, assuming a product/facility life of five years, the project in a world of no taxation (what a thought!) would probably not be undertaken because of its negative net present value of (385 000). However, the above analysis is far from being either complete or realistic. Not only should the impact of both taxation and inflation be taken into account, since they can have a substantial effect on the viability of the apparent investment, but also the derivation of the cashflows and the assumptions behind the timing of both the cashflows and the capital expenditures should be clarified.

For example, substantial cash-inflows and cash-outflows do not occur instantaneously, this being demonstrated by figure 3.7, for the development costs of an FMS. A similar, albeit more detailed analysis would need to be carried out for all conceivably related cash-inflows and cash-outflows.

These cashflow figures are likely to be derived from a number of sources, typically:

- Conceptual and Detailed Design costs
- Simulation costs
- Computer hardware and software costs
- Process equipment costs
- Material handling equipment costs
- Tooling costs
- Labour costs
- Recruitment and training costs
- Reductions in scrap
- Reductions in inventory
- Reduction in maintenance costs
- Cost of power
- Residual value of equipment

However, while the list is not complete, it is intended to contain the more important 'tangible costs and benefits'. In many respects these are the easiest parameters to obtain, verify and justify. However, as intimated above, many of the assumptions used to calculate the above figures could be highly relevant, for example, the allocation of general overheads etc. It is essential that all these assumptions are clearly identified within the analysis. Also the fact that there are a significant number of substantial, virtually unquantifiable benefits likely to be accrued as a direct result of the investment should be mentioned explicitly. All this data should be clearly presented in a standardised format, similar to that used for the component-related data prepared in chapters 4 and 5. Indeed, such a format will be used for the remaining analysis examples.

One quite popular method of ensuring that a high level of confidence exists in the results of the analysis is to subject the cashflows to a certain degree of risk. For example, the 650 000 expected during year 2 might have been as a result of a 50 per cent expectancy of 800 000 and a 25 per cent expectancy of 1 000 000. A variation of this technique is to add a 'risk premium' to the cost of obtaining the capital. The total capital amount could then be varied according to the different level of risk likely to be experienced with the alternatives being considered.

A further method of introducing a factor for uncertainty is to carry out a sensitivity analysis. This necessitates establishing the effect on the various results if all does not go according to the original plan. For example, the facility could be a year late, or early, in starting production. The sales

ALL−PARTS−MADE−FAST LTD,

Plant 3 − Flexible Manufacturing System Project

INVESTMENT APPRAISAL	Title: IA1	Ref to: IA/R1	Date:
			Author:

Net Present Value Calculations						Amounts in 000's (−)
Year	Cashflow before tax	Capital allowance (1,2)	Taxable returns	Tax due (3)	Tax paid (4)	Cashflow after tax
1	0	3,000	(3,000)	(1,500)	0	0
2	650		650	325	(1,500)	2,150
3	1,100		1,100	550	325	775
4	1,500		1,500	750	550	950
5	2,000		2,000	1,000	750	1,250
6	0		0		1,000	(1,000)
	5,250	3,000				4,125

ALL−PARTS−MADE−FAST LTD,

Plant 3 − Flexible Manufacturing System Project

RELATED DATA	Ref: IA/R1	Date:
		Author:

Title: NPV Calculations inc. Tax	Sheet ...... of ......

Comments: ....

....

Notes:

(1)   Assumes all allowances may be taken in first year
(2)   Assumes all capital expenditures are allowable
(3)   Assumes a 50% Corporation tax
(4)   Assumes tax payments are delayed by one year

General assumptions:

Data derived from Related Data sheets .....
Taxation levels obtained from .....

.... etc.

Figure 10.2   NPV calculations including taxation

forecasts might have been underestimated, or overestimated, by a certain percentage.

It is certainly highly desirable that these various types of sensitivity analysis are carried out, since if nothing else, they will clearly identify those elements of the project which exert proportionately the most influence. To perform the analysis correctly could involve a considerable amount of calculation, hence once again, the use of a computer-based analysis tool is highly recommended. This facilitates the comparison of alternative facilities in the light of such features as late delivery of equipment, market fluctuations, changes in company manufacturing strategy etc. For example, such an analysis might be invaluable as a demonstration of the ease with which a well-designed FMS can be adapted to produce different products, in comparison with a transfer line. In addition, if, say, the cost of the computer system is likely to vary by 10 per cent, then the effect of this on the total investment can be clearly demonstrated.

However, assuming the NPV analysis shown in figure 10.1 is carried out again, this time making an allowance for the effect of taxation, the values involved change quite significantly, as shown in figure 10.2. Additionally, the impact of inflation can be taken into account, thus diminishing the real value of the cash-inflows, as is shown in figure 10.3.

ALL—PARTS—MADE—FAST LTD,
Plant 3 — Flexible Manufacturing System Project

INVESTMENT APPRAISAL	Title: IA2	Ref to: IA/R2	Date:
			Author:

NPV Calculations inc. tax and inflation   Amounts in OOO's   (−)

Year	Cashflows after tax	Discount factor of 10% /annum	Cashflows after tax and inflation	Total Cash Outflow	NPV
1	O	0.909	O		
2	2,150	0.826	1,776		
3	775	0.751	582		
4	950	0.683	649		
5	1,250	0.621	776		
6	(1,000)	0.564	(564)		
	4,125		3,219	(3,000)	219

Figure 10.3   NPV calculations including taxation and inflation

A number of assumptions were made in order to carry out these two analyses, for example, the fact that cashflows may be discounted at the same rate of depreciation, when in practice differential rates would need to be applied to the various elements comprising the cashflows. Assumptions such as these should be clearly stated on cross-referenced Related Data sheets, as shown in the second half of figure 10.2.

However, although this analysis demonstrates the methodology of the financial justification of a major capital project, it remains a superficial example, and certainly does not contain all the detail which would normally be required. The interested reader is recommended to refer to one of the many texts devoted entirely to this topic (see References and Recommended Reading at the end of the book for some suggestions).

The following section is intended to show how some of the values used should be prepared and, perhaps more importantly, how they should be presented. It is a sad but true fact of life that all data produced needs to be not only perfectly consistent and arithmetically correct, but also presented in the correct manner.

10.5   Preparing for the analysis

The first step towards a comprehensive analysis will be the preparation of an appropriate database. This should be prepared in such a form that it is readily usable by one of the many microcomputer financial appraisal packages that are generally available. Each table or spreadsheet is likely to result in one or more entries within a later table. Therefore, if a microcomputer analysis package is being used, it is important that the structure of these tables is designed carefully, in order that they may be linked from the outset. This will ensure that whenever a change is made in one table, it is automatically reflected in all the related data tables. Also, it is sometimes of value to use a naming convention for these tables which gives an indication of the level of detail available. For example, 'level 1' data might be the most detailed information, while 'level 7' data might be the final summary analysis presented to the main-board. An example of what could be classified as 'level 4' data, namely a relatively high-level capital requirements' comparison of the four facility options which would typically face a company which is considering investing in FMS, is shown in figure 10.4.

Needless to say, more detailed sheets would need to be prepared for all the cash-inflows and cash-outflows relating to the project. For example, topics which would need to be considered would include:

ALL−PARTS−MADE−FAST LTD,
Plant 3 − Flexible Manufacturing System Project

INVESTMENT APPRAISAL (Level 4 Data)	Title: IA3	Ref to: IA/R3	Date: Author:

Capital requirements comparisons							Amounts in 000's (−)	
Expense type	Existing		Same new		CNC		FMS	
	Total	%	Total	%	Total	%	Total	%
Conceptual design	0	0	0	0	10	2	25	2
Detailed design	0	0	5	2	10	2	75	6
Computer hardware	0	0	0	0	5	1	75	6
Computer software	0	0	0	0	5	1	150	12
Process equipment	0	0	250	92	500	87	550	47
Material handling	0	0	0	0	0	0	250	22
Building renovation	10	100	20	4	10	2	25	2
Training	0	0	5	2	25	5	35	3
.... etc.								
TOTAL	10		280		565		1185	

ALL−PARTS−MADE−FAST LTD,
Plant 3 − Flexible Manufacturing System Project

RELATED DATA	Ref: IA/R3	Date: Author:

Title: Comparisons of capital needs   Sheet ..... of .....

Comments: ....

....

Notes:

(1)   Effect of inflation is not included
(2)   Regional development grants are included (refer to IA/15)
(3)   Technology development grants are included (refer to IA/16)
(4)   Timing of expenditures is as shown in IA/17
(5)   Conceptual Design costs are shown in IA/21
(6)   Computer hardware requirements are detailed in IA/22
(7)   Building costs (ref. IA/23) include loss of production
(8)   Start−up costs (ref. IA/24) include scrap and launch

.... etc.

Figure 10.4   Expansion options capital requirement comparison

(1) *Start-up costs*

- Installation
- Commissioning
- Training and recruitment
- Part program preparation
    . . . etc.

(2) *Annual operating costs and savings*

- Operators and supervisors
- Maintenance and rates
- Power
- Consumable tooling and coolant
- Work in progress, swarf reclamation
    . . . etc.

(3) *Once-off costs and savings*

- Regional development grants
- Technology development grants
- Inventory
- Learning curve
- Publicity
    . . . etc.

Following the collection of all this data, and hopefully its entry into an appropriate microcomputer spreadsheet, it is then appropriate to carry out a sensitivity analysis. Ideally, this should be started at the lowest level of data, with the 'expected' sums now being replaced by estimates of the best-case and worst-case possibilities. Once all the necessary data has been prepared (and the above list of tables and entries is by no means complete) it will then be necessary to take into account such factors as inflation, the timing of expenditures, taxation etc. The end result would be the preparation of a sensitivity analysis of all the various investment alternatives. The overall effect of these changes on the profitability conclusions of the initial analysis should then be considered in detail.

10.6 Presentation of the results

Ultimately, this is the data which will have to be used by those responsible for giving the project formal approval. It is therefore imperative that the data is presented in a clear, concise, uncluttered manner, and, needless to say, is arithmetically correct. Certainly, the use of different type faces (italics, bold etc.) as an aid to legibility is to be thoroughly recommended.

The end result would eventually be a summary of the analysis, perhaps supported by some Related Data sheets. Since upper management is unlikely to have the time, or the enthusiasm, to read through a substantial quantity of numerical analysis, the Project Team should resign itself to the fact that the less data that is requested, the better the analysis. The financial summary which is presented to the main-board should certainly be no longer than a few pages, though this could perhaps be accompanied by a more lengthy document. Figure 10.5 shows two data sheets: an overall summary in a relatively acceptable format, and a Related Data sheet which is too vague. Ideally, if an assumption is of sufficient importance to be mentioned in the summary, then the assumption itself should be included within the Related Data sheet; it should not be necessary to trace through all the references in the supporting data.

It cannot be overstressed how important these summary pages are likely to be to the credibility of the whole proposal. If the detailed financial analysis shows the project benefits to be marginal, then the decision as to whether or not to proceed with the project will almost certainly rest on the difficult-to-quantify, intangible benefits. Often a plausible attempt at quantifying these benefits, albeit with a sensitivity analysis, can aid the appreciation of their potential magnitude, however, care should be taken to ensure that the conclusions are not misleading.

It is worth adding that a picture is indeed worth a thousand words! If the presentation of the data can readily be made in a pictorial or graphical form, then this will certainly improve its comprehensibility, and maintain the interest of the reader.

10.7 Concluding comments

Unfortunately, as mentioned previously, it is too often the case that the Project Team becomes too involved in the facility it is preparing. One wonders how many times a justifiable proposal has been rejected either because too little thought was given to the preparation of the financial analysis, or because the engineers involved were too technically orientated, or because company accountants and management had an inadequate comprehension of the technological and marketing issues involved. Either way, correct emphasis should be given to the 'middle line' benefits, as they are sometimes called — for example, customer goodwill, market positioning, market pro-portion growth potential etc. It is said that one of the reasons that the

ALL−PARTS−MADE−FAST LTD,
Plant 3 − Flexible Manufacturing System Project

| INVESTMENT APPRAISAL (Level 7 Data) | Title: IA5 | Ref to: IA/R7 | Date: |
| | | | Author: |

| Investment options − Summary comparison | Amounts in 000's   (−) |

	Existing	Same new	CNC	FMS
Capital costs	10	270	565	1185
Less grants	0	0	166	328
Investment	10	270	399	857
Av. costs/yr				
Av. earnings/yr				
Once−off costs	etc.			
Once−off savings				
NPV				
IRR				
Hidden benefits	*0*	*0*	*Some*	*Many*

ALL−PARTS−MADE−FAST LTD,
Plant 3 − Flexible Manufacturing System Project

| RELATED DATA | Ref: IA/R7 | Date: |
| | | Author: |

| Title: Summary comparison support | Sheet ...... of ...... |

Comments:

Notes:

(1) Capital costs are as shown in IA/11−14
(2) Grants are as shown in IA/15−16
(3) Sensitivity analysis results are shown in IA/31−34
(4) Once−off costs and savings are shown in IA/26−27
(5) Implications on personnel reductions are shown in IA/37
(6) Inflation and taxation factors supplied by ...
(7) NPV and IRR calculations are shown in IA/1
(8) Hidden cost and benefits are shown in IA/R42

.... etc.

Figure 10.5   Facility options — summary investment comparison

Japanese have been so successful recently is because they do indeed give reasonable weight to these factors. In particular, the need to grow a market and to increase the proportion which is owned, is a major consideration which is often neglected. Finally, the presentation of the data is fundamentally important. A good argument, badly presented, is doomed to failure.

11 Installation and Commissioning

11.1 Introduction

By the time one reaches the installation stage of an FMS project, all the major design issues regarding the system should have been resolved. All that remains will be to prepare a suitable site for the FMS, and embark on the commissioning process. This certainly sounds simple enough, but in fact there are a number of important issues which need to be addressed during this phase of the project. During this chapter, a number of these more practical issues will be discussed.

11.2 Background comments

Probably the single most important factor to be borne in mind when considering the installation and commissioning of an FMS is that this process is likely to be both lengthy and resource-intensive. Undeniably, it is extremely difficult to estimate how long commissioning should take. As might be expected, it rather depends on the inherent complexity of the system and on how many problems are exposed during early testing, and once discovered, how easy they are to correct. It is during this phase of the project that one finds out how comprehensive the Detailed Design of the system really is.

Certainly, it is nearly always the case that insufficient time is allowed for debugging the system. It should be appreciated that there are a number of distinct phases to this complete process, all of which have to be carried out in a carefully controlled manner, first on integrated subsystems of the FMS and, finally, simultaneously on the whole system. Hurrying this phase of the project is false economy since problems found once the system is in production are likely to be significantly more difficult to rectify.

Theoretically, the physical installation of the system should be relatively straightforward, especially if the whole process has been well planned.

However, there are bound to be some unplanned variations in the equipment deliveries. At times such as these, a computerised project planning tool, as introduced in chapter 3, is extremely helpful, especially if, for example, equipment has to be installed in a particular order.

11.3 FMS siting implications

The architectural requirements of FMS are relatively straightforward, not differing much from those of the individual equipment located collectively. There used to be a tendency to install such systems in new buildings. There is no need for this, but if the opportunity exists, the possibility of locating the FMS within a carefully planned, pleasant working environment is an appealing possibility.

If an existing building is to be used, then care should be taken to ensure that this represents both a feasible and an economic course of action, in both the long term and the short term. In the short term, accommodating the FMS within an appropriate existing, working environment can be less expensive. However, in the long term, taking into account the expansion requirements of the system, and its integration with other parts of the company's present and proposed manufacturing systems, a new facility now might represent a less-expensive course of action.

When planning the installation of an FMS, the first factor to be considered is the floor space required for the equipment. The overall layout of the FMS will probably have been derived as a result of computer simulation. The implications of this layout will then have to be considered in great detail with regard to the requirements imposed on the building where the FMS is to be housed. For example, there could be a requirement for a trench for a flume coolant removal system. Similarly, special foundations for either or both of the process equipment and the material handling equipment may be needed. Also, there will be a need to ensure that services (power, air etc.) are made available in the correct locations; these will need to be laid out taking into consideration such factors as the need to separate power lines from data cables. All these, and many other requirements, will have to be accommodated. In the past, there has been a tendency to neglect some of these details, in particular the appropriate separation of cables, but at best this results in unreliablity being designed into the system from the outset.

Obviously it is desirable to have all the building works carried out at a time when, even allowing for slippage, the work will be complete before any equipment is due to arrive. This would start with major building works and be followed by such tasks as redecoration. As with the development of the FMS itself, a project plan should be created for this task, ideally on a computer. This would assist with the subdivision and scheduling of tasks, and also the impact of any changes made to the building requirements.

These tasks are likely to be relatively straightforward if an entirely new building is being used. Similarly, if an existing facility is being utilised, with perhaps the current equipment being relocated to a new area. However, if the FMS has to be installed progressively, in the midst of an existing production facility, and this has to be able to continue manufacturing products while the FMS is being installed, then the whole installation process becomes somewhat more complex. Indeed, this situation could well be further aggravated if some of the equipment which is currently being used for production is eventually going to be used within the FMS, and this situation is more likely, since complete green-field FMS installations are less frequent.

In fact, there are essentially three ways in which FMS may be implemented:

(1) 'Green-field' site.
(2) 'Linked islands' of automation.
(3) The evolutionary approach.

The 'green-field' site is when the entire FMS — building, equipment etc. — is new. This type of project represents an exciting development opportunity for all involved. Unfortunately, although in all respects it is probably the most attractive FMS option, it is usually also the most expensive. However, the experience which can be gained by enthusiastic members of the Project Team during such a project can be quite unique.

The 'linked islands' approach is probably the most practical method of progressing towards a highly automated/integrated manufacturing facility. Frequently, it is not possible to finance all the investment necessary to progress from a traditional manufacturing environment directly to an up-to-date highly automated environment. Indeed,it is rarely desirable that such a transition should even be attempted, since the strain placed on both the shop-floor personnel and production control systems might be quite unreasonable.

Providing the long-term plan is well thought out and clearly defined, it is quite feasible, and indeed in many respects highly desirable, that automation is installed in islands, which later may be linked together to form a larger integrated manufacturing system. This not only provides the opportunity for the earlier elements of the automation to help fund the later elements, but also means that the changes in the level of automation technology being applied on the shop-floor can be contained within manageable steps. This is the type of approach which appears to be becoming more and more popular. However, if this is the chosen route, it is imperative that the long-term strategy is thought out at the start in considerable detail.

The final method of installing this type of automation equipment is essentially a subset of the linked islands of automation, namely the 'evolutionary approach'. In this method, plant and equipment is purchased on a piecemeal basis. If the long-term plan is good, then eventually islands of automation will be formed. With this route to automation, taking account of the rate of technology change becomes more important. For example, is it feasible and desirable to integrate the equipment of today with the equipment of tomorrow? At least if one has an island of automation, any problems are localised and hence do not affect the overall operation of the factory. The feasibility of such an approach is becoming clearer with the rising popularity of MAP (see chapter 12) as a generally accepted standard for factory-wide communications.

However, one of the consequences of both these latter approaches is that eventually the FMS will be installed within an existing facility. This could occur either progressively or *en masse*, but during this time it is likely that the traditional production system will need to continue to operate while the new equipment is being installed. In fact, the existing equipment might even have to be worked harder to make up for the inconvenience caused by the FMS installation process. This naturally poses considerable incremental burdens on the FMS installation and commissioning processes, and might even influence the layout of the FMS. For example, it might prove undesirable to move an item of equipment which is of key importance to the remainder of the existing production system.

11.4   Planning the FMS installation

Regardless of whether the FMS is being installed within an existing facility or in an entirely new building, one of the main problems will be to ensure that the necessary work is carried out in the correct order and in a timely manner; not compromising any of the equipment deliveries, which themselves will probably need to be carefully scheduled to fit in with the overall installation plan. This will be particularly important if there is little or no cranage available within the building, since this implies that any heavy equipment will have to be manually handled into position, perhaps on rollers with the aid of probably a fork-lift truck.

This problem will be further exacerbated if insufficient space exists to move equipment about freely once it has been moved into the building. If, for example, the FMS is to be installed in a relatively narrow building, perhaps with a material handling system running along the middle, it might well be that the process equipment can only be installed in a particular order (depending on the location of the doors to the building). Assuming all the equipment arrives on schedule this might not present any problem, but if, as is likely, some items of equipment are late, this may cause a considerable delay to the installation and commissioning of the equipment, and hence to the system as a whole.

Once again, the use of a computer aided project planning system can vastly reduce the impact of last minute changes such as these by allowing the Project

Manager almost immediately to assess the impact of the change and the impact of the various courses of action open.

Interestingly, for some reason these project planning packages do not appear to have been particularly popular among manufacturing system engineers. Presumably because only recently have projects as complex as FMS become relatively frequent events. Certainly, these packages have been used for some time, and very successfully within the civil engineering environment. Their lack of popularity within automation-related environments might well be indicative of a general weakness in production engineering capability.

The order in which work is carried out when installing an FMS will vary considerably from one system to another, not only for the reasons stated above, but also because of the wide variety of equipment which could be required. For example, in an assembly FMS which might only be part automated, much of the equipment might not be particularly heavy. Therefore, the probability of substantial installation problems being encountered is considerably reduced. However if the FMS is, say, machining-centre based, perhaps with an AGVS or a rail-guided transport system, the situation is unlikely to be so straightforward.

Although it is difficult to generalise, the order in which the major tasks would typically be carried out is essentially as follows:

(1) Prepare building (generate design specification, bill of materials, schedules etc.).
(2) Accommodate specific equipment requirements.
(3) Install equipment.
(4) Connect equipment.
(5) Commission equipment/systems in isolation.
(6) Commission equipment/systems collectively.

The preparation of the building and the FMS floor area is relatively straightforward, especially since much of this work can be carried out after the FMS detailed design has been completed and probably long before any of the equipment is likely to arrive. Some of the more obvious requirements of this phase of the installation are ensuring that the floor is appropriately level and that suitable foundations exist where required. This would probably include reinforced sections of floor for particularly heavy equipment and vibration absorbing mounts for co-ordinate measuring machines and other delicate machinery.

Once the locations of the major items of equipment have been fixed, it will be necessary to ensure that ducts are correctly positioned both in the floor and maybe trunking in the roof space for the laying of services and data cables repsectively. It should be noted that it is becoming quite common to feed as many of the services as possible from the roof space to the equipment since this obviates the need to dig trenches in the floor and makes it easier to, for example, lay tracking wires for an AGVS, not to mention changing the layout of the FMS (which could well be a particularly important factor if the FMS is to be installed as part of a long-term plan). Of course, the disadvantage is that such cables might obstruct any overhead equipment, such as mobile cranes etc.

11.5 The working environment

Certainly, during such major building operations the opportunity should be taken to make the new working environment as pleasant as possible. Both the building and the equipment should be in pleasant and practical colours. The colours could even be used to indicate the functionality of the equipment. Also, it is desirable to ensure that both the equipment and the environment generally are kept as clean as possible. Sometimes this is not easy, but if achieved, the overall effect is beneficial for all concerned.

If possible, as much natural light as is reasonable should be allowed to illuminate the work space. If vision systems and other light-sensitive equipment are to be used within the FMS, then care should be taken to ensure that the lighting conditions, especially in the vicinity of this equipment, remain constant, and without glare. For example, it is obviously unsatisfactory if a vision system is installed in such a manner that it operates correctly when the sun is obscured by clouds but not when direct sunlight illuminates the facility, perhaps through a skylight. Suffice to say that this requirement should not be used as sole justification to compromise the rest of the working environment. But it is possible that vision equipment etc. will need to be fairly well shrouded to function consistently if the ambient lighting conditions are subject to substantial variation.

If the FMS is to be installed within an entirely new building, it might well prove advantageous to consider the incorporation of a suspended roof in the design. These are now quite proven structurally and offer the considerable advantage of not obstructing the factory floor space with a plethora of supporting columns.

If the FMS is to be installed within an existing facility, the need for careful planning is even more apparent, particularly if production is to continue while the new system is being installed. Suffice to say that the Project Manager's task is going to be quite difficult enough at this time with all the various suppliers' efforts coming to fruition, without having an irate Production Manager finding difficulty in maintaining the expected output levels from the existing equipment.

One of the most difficult problems with installing an FMS within an existing facility arises when old equipment has to be moved out or simply relocated when new equipment is being moved in. It is during these times that most disruption is likely to be caused.

Once the old equipment or any other obstructions have been removed, preparation of the floor and walls can start. This could mean digging foundations and trenches where needed. The floor should eventually be finished to a good flatness, sealed and, ideally, with the above comments in mind, painted. This is not only for the benefit of the process equipment, but also to facilitate both the installation and eventual operation of the material handling systems.

Relatively light devices, such as robots, which usually do not require special foundations as might be needed by a heavy and accurate machine-tool, can be moved into position as soon as any associated equipment has been installed. Possibly, the commissioning processes of items of associated equipment can be overlapped to advantage.

However, with robots and any other devices which interact with multiple items of equipment, care should be taken to ensure that an appropriate location is provided. This need not necessitate a great depth of concrete, but does justify an adequate means of securing the equipment to the floor — for example, long enough bolts etc. For instance, if a robot accidentally collides with, say, a machine-tool, one does not want the robot to move from its location. This would necessitate the reteaching of all the robot programs previously written for that robot. However, the use of fixed datums can reduce the potential severity of this problem.

After the pneumatic drills have finished their work, the dust has settled and everyone can hear themselves think again, one can pass to what in many respects is the most peaceful and rewarding phase of an FMS project, namely finishing the building prior to the arrival of equipment. This effort would probably be concluded by the installation of an appropriate air conditioning or extraction system. This will be particularly important if the process equipment is either delicate or able to generate a substantial amount of either heat or fumes. It also helps keep the environment pleasant for the few operators that are present.

11.6   Installing equipment

With the building completely prepared, the next step is to receive the first items of equipment. Each would be moved into its appropriate location and connected up to its respective services. With the equipment ready for operation in isolation, it should be fully commissioned. Little consideration should be given at this time to how the equipment interacts with the rest of the FMS. There should be plenty of time to address these issues later in the commissioning process. A fundamental requirement for all the equipment to be used within the FMS is — first commission it in isolation. Only then should one progress to the next phase.

To emphasise this point, where a number of items of equipment closely interact, as for example is often the case with a robot and a machine-tool, both items should be commissioned and fully tested in isolation before they are tested together. Where ancillary equipment such as a flume swarf removal system or a conveyor is involved, these too should be commissioned in isolation, before being integrated and commissioned collectively. However, items such as these should be considered part of the basic machine-tool subsystem. Therefore the ancillary equipment should be commissioned prior to, for example, the machine-tool being commissioned with the attendant robot.

Once all the equipment is operating correctly, both in isolation and as integrated production subsystems, the next stage of commissioning would be to test the communications systems. These tasks are likely to be somewhat complicated by the fact that the control software is unlikely to be complete at this point. However, the detailed design of the software should have taken into account the commissioning schedule for the FMS, and therefore at least the basic routines necessary to check out the equipment within the FMS should be available. Typically these would comprise the ability to send and receive part programs, and perhaps set-up files; the ability to monitor status changes in the equipment from the FMS control system; and the ability to exert control over the equipment from the FMS control system.

The communications between the control computer and each individual item of equipment should again be checked first in isolation. This will ensure that any problems which are encountered can be traced back rapidly to their source. If more than one function of the system is tested at once, faults may be due to either one of the two systems being tested, or indeed the combination of the two systems. Suffice to say, it is not unusual for equipment which is known to have been operating correctly in isolation suddenly to start failing when operated with other equipment. If, perhaps in an attempt to save time, the equipment is commissioned collectively from the start, when faults do occur they are far more difficult to trace.

However, because of the inherent complexity of most flexible manufacturing installations, it is unlikely to prove possible to test every conceivable operational combination of every piece of equipment. This would probably represent a task which would take many years to complete. Nevertheless, the major combinations should all be systematically checked. How these major interactions are selected is likely to vary substantially from one environment to another. However, they would certainly include batch changeovers, quality control actions, recovery situations etc. It is particularly important that this process is carried out correctly at this time, since there is unlikely to be another opportunity to carry out the work in such a controlled manner.

11.7 Safety equipment

With all the equipment known to be operational, it is important that appropriate safety equipment is installed. Although production equipment tends to be provided with safety guarding by the original manufacturer (as, for example, is the case with most machine-tools), much equipment cannot be guarded satisfactorily in this way since guarding requirements are entirely dictated by the particular FMS within which the equipment is located (as, for example, is the case with most robots). Frequently, additional guarding will be required because of the combinations of equipment being used within the FMS.

This additional guarding may well be supplied by one or more of the equipment manufacturers. If this is the case, care should be taken to ensure that wherever possible it is of one standard design and especially of one physical appearance, even though different areas within the FMS might have their function distinguished by floor colour. The design can be established during the specification and design phases of the FMS. It is essential that whatever action is required to be taken by an operator in one part of the FMS to make equipment safe, is as near as possible to the actions which need to be taken in other parts of the system.

Certainly, the guarding of automated equipment is an often neglected topic. The remainder of this section of the chapter deals in some length with this subject to show the variety of approaches which could be adopted. The section should also be seen as an indication of the depth to which all such topics should be considered prior to a particular approach being selected.

Without doubt, there are many ways in which poorly designed guards could actually be hazardous. Despite the fact that 'non-physical' (sometimes called 'non-contact') methods, such as multiple infra red sensors, have grown in popularity, it is worth bearing in mind that the traditional approach of physical guarding, when correctly designed, remains efficient, cheap and highly effective.

Optical sensors have their advantages in certain situations, particularly, for example, where the equipment is not entirely automated, and hence frequent manual interaction is necessary during or even as a part of the manufacturing process, and especially if the manual interaction needs to be relatively fast. In this type of situation it will probably prove impractical to have, say, a metal guard moving into a position where it will ensure the operator is clear of the equipment. Therefore an optical guarding system, with no moving parts and their associated inertias, can be extremely useful. A disadvantage is that since the guard is not actually visible itself, it might well be triggered more frequently than would have been the case with a physical guard. However, a good overall design can minimise the impact of this problem.

For groups of equipment, where there might be a machine-tool and a robot, perhaps individually or collectively interfaced to a materials handling system, optical guards are likely to prove impractical, as well as prohibitively expensive, because of the large area which might need to be protected and its probably irregular shape. In these situations, physical guards do have some significant advantages as far as operator safety is concerned, since they are, for example, easier to see and less expensive to install.

Although well-designed physical barriers are excellent in performing their basic function, namely keeping people out and both equipment and components in, they are possibly a restricting factor when access to the equipment is essential. To some extent this is bound to be the case, but again, with a good basic design, the impact of this drawback can be minimised.

The three key requirements of guarding within a highly automated facility will now be considered in turn.

(1) To keep people out when the operating mode of the equipment is potentially hazardous.

This is usually why the guards are purchased in the first place, either on the instructions of the user or of the supplier, or perhaps because they are necessary for the system to comply with health and safety regulations. Frequently, it is the only reason why people feel it is necessary to have guards at all.

However, it should be appreciated that the guards must prevent any unauthorised and/or untimely access. This not only includes operators but also visitors. If the FMS is both advanced and successful, substantial numbers of visitors can be expected and therefore appropriate precautions, such as raised walk-ways and eye protectors, must be included within the specification of the system.

However, it will probably be useful if a reasonable visual inspection of the equipment that is being guarded can be made by someone even when the equipment is in operation. This can quite easily be achieved by selecting guarding that is either all or part made of a transparent material. For example, if the equipment is not too powerful, guards manufactured from Perspex or an equivalent material could suffice, perhaps with a metal supporting frame. Alternatively, some form of metal guarding could be used, perhaps with sheet metal for the lower half and steel mesh for the upper half. Such guarding not only provides the rigidity and strength needed, but also the means of effectively enclosing the equipment when it is in operation.

Whichever type of guarding is selected, it should be high enough to make unauthorised entry extremely difficult. This is not only to ensure that people will not climb over it but also to ensure that anything ejected from the equipment will not endanger anyone standing nearby.

On the other hand, it should be relatively easy to remove the guards when the equipment requires major maintenance, or perhaps has to be relocated. This maintenance facility can be provided in a variety of ways. For example, by having the guards manufactured in relatively easy-to-handle sections, all of which can be located relatively easily, on side supports. These can themselves be located on rigid floor-mounted (or wall-mounted etc.) supports. This ensures that the guards are rigid when required, and yet can be disassembled relatively easily if, for example, substantial, and yet temporary, access is required to the equipment.

(2) Keep equipment and components etc. within safe confines.

This is a use for guarding with which only physical guards can easily comply. It is frequently forgotten that machine-tool guards are not only intended to keep people out of the machine, but more importantly to keep swarf, coolant and fumes etc. within the confines of the machine environment. For example, guards ensure that any components which perhaps have not been clamped properly, do not have sufficient energy to be thrown outside the machine after they have been released mid-cycle. This problem is tending to increase in its severity as the cutting tools for machines (such as ceramics etc.) necessitate much higher cutting speeds. This requirement to keep components and equipment within the safety fencing applies equally well to robots which are not usually supplied with their own guards, as well as to the guards around machines, even if this necessitates the ordering of non-standard or even additional equipment to the manufacturer's standard guards.

Not only should the guarding be strong enough to restrain an impacting component, regardless as to the type of equipment from which it originated, but also it should be able to contain an item of equipment such as a robot which, when travelling at maximum speed, may strike the guards at their weakest point. Hopefully such circumstances should not occur but, if they did, then anybody standing near the guards should be safe. Needless to say, testing that guards comply with these requirements is likely to prove quite exciting. It is not often one has the opportunity to program a robot to rapid-traverse into a section of guarding! Such tests also give an indication of the robustness of the robot and its location, and whether or not it will be necessary to reprogram the robot after a severe impact.

As a further consideration, it might well be appropriate to ensure that the guarding does not permit fumes from the manufacturing process to pervade the factory atmosphere. The requirements of this type of guard are essentially as previously described, but with the added complication that they might have to be air-tight and fitted with extraction equipment to help maintain a pressure slightly lower than atmospheric within the confines of the guards. As expectations mount about the desirability of pleasant and safe factory environments, considerations such as these are likely to become even more important, especially as some of the coolants and other chemicals used within various manufacturing processes can be harmful if inhaled. Alternatively, if leakages or simply a build-up of condensed fumes occur, equally hazardous conditions could be generated for the operators.

Another feature of guards is the need to:

(3) Permit occasional access.

Having stated that guards must be strong and yet relatively easy to move, there is also the intermediate case which needs to be catered for, namely that of operators requiring occasional access to the equipment etc. There are essentially two types of access which typically fall into this category, namely 'mid-cycle access' and 'end of batch access'.

Mid-cycle access might be required for manual inspection of a component, or perhaps to recover and restart the equipment after it has failed. The main difference between these two types of event is that inspections etc. are planned and probably occur relatively frequently (depending on the level of control which is being exerted on the process itself), while recoveries of equipment will occur at random.

For inspections etc., probably the easiest way of obtaining the required component for the gauging or quality control checks is to have the component automatically placed at a location where an operator can easily and safely reach it, assuming that, for the purposes of this discussion, it is highly unlikely that it would be acceptable to stop production during these checks.

In some environments the contrary could be true, however, in control terms these are purely a subset of the above. The main difference is that to enable the control system of the FMS to keep track of the components between inspections would inevitably be more complex if the production process is allowed to continue while a component is waiting or in the process of being inspected. Having a component location point close to a sliding window would be one example of how this could be achieved. Sensors could perhaps be mounted close to the window to ensure that the equipment could not gain access to that area while it was being used by an operator. A more expensive, but possibly safer, method would be to have the part automatically removed from the reach of the robot and presented to the operator. This could be achieved by a relatively simple conveyor mechanism, though some component-specific fixturing might be required.

To recover equipment after a failure of some description might well require quite a substantial amount of access. For example, a robot may have been wrongly programmed so that when a certain set of circumstances arose the robot collided with a machine, dropped a part or perhaps the robot's gripper mechanism became damaged. Typically, during these incidents an operator will be required to spend a considerable amount of time (the less the better, and the equipment should be designed to ensure that this is the case) with the robot, ensuring that it is ready to produce the new component or product.

However, whether the access is required for the changing of robot grippers to hold a new component (that is, a set-up operation) or because the original gripper has been damaged makes little difference to the FMS control system. In either case it may be necessary to carry out a considerable amount of work on the production equipment which is being guarded. To provide safe access for an operator in these circumstances can represent quite a complex problem, since it is quite likely that the equipment will have to be operated, in albeit a slightly detuned manner, before the operator will be satisfied that the equipment is ready to produce the new components. It is an unfortunate fact that until true off-line programming becomes viable (see chapter 12), it will be necessary for operators to be physically in close proximity to the equipment which is being adjusted. This, potentially, exposes the operators to a certain amount of risk which, although to some degree is currently (and for the foreseeable future is likely to remain) unavoidable, can nevertheless be minimised by good guarding.

Typically, devices such as robots (and these will be used as the prime example of a device which on the one hand must be adequately guarded but on the other must be provided with the means for occasional operator access) need to be quite well guarded because they move quickly within quite large areas of space around the equipment to which they are interfaced. Of course, one of the most insidious attributes of this type of equipment is that it moves quietly and with little or no warning.

Occasional safe access to the equipment can be provided by the obvious, namely a door, but the operation of this must inhibit further activity of the robot. What is more, it should not be possible to use the access provided by the door incorrectly. Hopefully it would only ever be used by appropriately skilled personnel, but in the event of this not being the case, it should be as difficult as possible to unwittingly become involved in an accident.

A typical means of achieving this is to have the door operated by a removable key; this ensures that only those who have been allocated keys are able to obtain access through the guards. Of course, the operators will complain that there never seems to be a key when one is required, but all the locks can be made the same, and individual keys worn around a belt etc. To ensure that keys are not left in the door, it should not be possible to restart the robot with the key in the lock. Once the door has been opened, the robot should be inhibited from further activity (and this should be achieved in a hard-wired form rather than by software). To provide the facility to operate the robot with the door open, there should be a second key location within the guarded area. When the key is moved into this location, it should be possible to operate the equipment only in a slow 'jog' mode. Suffice to say, it should not be possible to operate the robot at a fast speed if the door is not closed, and this should also not be permitted if the key is in the inside location. In addition, it should only be possible to close the door in the guards from the outside, and not if there is a key in the inside location. This will ensure that the door (which itself should always be designed to open outwards) cannot close and trap an operator within the guarded area.

Also, care should be taken to ensure that there are no 'pinch points' within the guarded area which could conceivably trap an operator between a rigid fixture and the robotic equipment. This is not always easy to do, especially if, as is usually the case, space is at a premium. However, it is essential that in the unlikely event of a robot moving quickly with an operator in the way, there is always somewhere for the operator to move where the robot can be avoided. It may be that the operator is pushed out of the way by the robot, but this is preferable to being crushed against a safety guard.

One relatively new method of helping to ensure that such an accident does not occur is to use pressure-sensitive mats around the robots. These are secured to the floor, and ideally, used in addition to some form of physical guard. When the robot is active, any one stepping on the mat automatically causes the robot to stop. These mats are very effective, though possibly somewhat expensive for the typical metal cutting factory (and one has to wonder how durable they would be in this type of potentially quite difficult environment). Within a laboratory or perhaps relatively clean assembly environment, however, they are ideal — though to date their application does not appear to have been particularly popular outside the USA.

Another aspect of operator safety which should be given careful consideration is that of how an operator should attract attention when in danger. Eventually the equipment itself will indicate that a problem has developed (perhaps a drive overloaded during the attempt to push an operator through the safety guarding!) by switching on, say, the flashing red bay attention light, which would need to be mounted in such a position that it is easy to see. Or perhaps by lighting an appropriate icon in the control room mimic display. Both of these actions are likely to be too slow if remedial action is to be taken by the other operators.

Ideally, the operators should carry some means of quickly and easily attracting attention. For example, they could be issued with what are sometimes called personal alarms — that is, small canisters of compressed gas which are capable of emitting a loud screeching sound when activated. Alternatively, hand-held radios could be issued, though it is debatable whether these would be of much use in a real emergency.

Certainly, it is important that whenever it is necessary for an operator to work near a particular piece of equipment, there is always an E-stop button within reach which will immediately halt the offending equipment. This is not always quite so easy to facilitate, since much equipment will often be working together in closely integrated units, and some items might be more dangerous when stopped quickly than if allowed to slow down gradually (for example, many grinding machines). Similarly, it may not always be desirable to shut down a complete system purely because there is a problem which may or may not involve an operator in one section. A multi-lane conveyor system is a good example of such a situation. If a pallet was to become jammed in a gating mechanism, it would probably only be necessary to close down the relevant gate and lane. Indeed, the whole issue of E-stops is nowhere near as straightforward as it might appear and as such should be given very careful consideration, ideally well before the installation and commisioning phases of the FMS. It is a sobering thought that someone's life might well depend on how well this and other aspects of the system design have been thought out.

Having said this, it follows that it is not desirable that only one operator should be responsible for the FMS at any time. At least one helper should always be available. Then if one operator is carrying out a task on the system and runs into difficulty, there is at least one other person who would quickly be able to provide assistance.

Also, the lighting system must be effective. With much equipment typically being packed into a relatively small floor area within an FMS it is possible to create an environment, especially at night if the system is to be run 24 hours a day, in which it will be difficult to see clearly while working in all parts of the system. As highly automated systems become more common, this problem is likely to diminish, since operators will be shared among several systems.

Finally, to assist with the safe operation of highly automated manufacturing systems, television cameras are becoming quite popular. These surveillance systems are now relatively cheap to install and maintain, and may be used both to monitor equipment and operators within the system. In one system within the USA, four such surveillance cameras monitor strategic equipment in each subdivision of the FMS. These produce displays at both the machines and in the central control room, the latter being equipped with a duplicate set of machine-tool controls which allows the machine parameters to be adjusted remotely. This is certainly both an effective and impressive installation, but probably prohibitively costly for the typical metal cutting factory.

11.8 Integrating the equipment and the systems

Returning to the commissioning process: by this time all the process equipment will have been installed and tested, both in isolation and as integrated subsystems, and also the communications to this equipment will have been proven. The next stage of the commissioning process is that of linking the process equipment with the main material handling system.

As with the process equipment, the material handling systems should be commissioned in isolation before being integrated with any ancillary equipment. Only after this procedure is complete should they be integrated with the rest of the FMS subsystems. First one or two process subsystems should be included with the main material handling system, and gradually more would be added as confidence in the operation of the system grows.

Depending on the inherent speed of operation of the material handling system, this phase of the commissioning process could be quite time-consuming. In order to test out any systems comprehensively, various sets of circumstances need to be created, initially for one of the FMS subsystems, then for a group of subsystems and finally for the entire FMS. If the cycle time of any of these items of equipment or subsystems is relatively slow (as will often be the case with a conveyor or an AGV), then even creating the test situations will be a lengthy operation.

After the process equipment, material handling equipment and communications systems have all been thoroughly tested, all that remains is the testing of the computer systems. For anyone who has little or no prior experience of complex computerised systems, this can be a particularly frustrating time for a variety of reasons. Not least of these is the fact that all of the equipment, ostensibly capable of producing good components, has been installed and fully commissioned in at least the traditional sense of the words, and yet none of the equipment is apparently available to assist with the production requirements of the company. In reality, the FMS would probably prove to be an admirable demonstration of how easy it is to produce scrap in large volumes. However, as mentioned earlier, it is imperative that this final phase of the commissioning process is not rushed, regardless of the pressures.

This phase essentially comprises the creation of major events within the system and the monitoring of the FMS's programmed response. In certain instances it will be possible to merge tests and hence reduce the period of time needed to complete the tests, but frequently this will not be either possible or desirable and the whole FMS will have to be cleaned out to enable a 'fresh' start to be made for the next test procedure. Suffice to say, this whole process is likely to take quite a considerable period of time, probably longer than everybody's estimates, but on no account should the process either be rushed or prematurely curtailed. Once the FMS has been turned over to production it will be extremely difficult to carry out well-prepared and documented tests to ensure that the software is functioning correctly, or even adequately.

All the software will probably not be delivered to the FMS at one time; like many other major elements of the system, it will probably arrive in phases. It is obviously important that the software deliveries are well co-ordinated with the arrival and testing schedules of other equipment. Probably the first element of software that will be needed for the commissioning of the FMS as a whole is that responsible for communications. To make this element function adequately for the testing procedures it is likely that some of the supporting software will also be required, for example, the creating of files for process equipment part programs and the monitoring of status changes within the equipment. Until the major part of the FMS has been commissioned, it is unlikely that much more of the software will actually be needed.

Once the production equipment has been fully tested and the communications links are all fully proven, more of the control software will be required. Particularly important will be the modules responsible for work allocation or scheduling, co-ordination of work transport and recovery, all of which will also require rigorous testing.

11.9 Training and recruitment

Another important aspect of implementing an FMS which as yet has not been mentioned is that of operator training. The installation and commissioning process provides an excellent opportunity to carry out this task, the fundamental importance of which is often overlooked. Ultimately, there is not much point in having a well-designed, debugged FMS if there is no one to operate it.

Ideally, the training process should be planned to be completed during the final stages of commissioning. This should then ensure that the learning phase is as short as possible and is nearly finished when production starts.

The training process itself will comprise a number of elements. Obviously, the operators will have to become proficient in the operation of the equipment and this may or may not include the ability to generate part programs. This will largely depend on whether it is company policy to have part programs written or amended on the shop-floor by the process equipment operator. Even if programs are not to be written on the shop-floor, it is likely that the imparting of some knowledge on this subject to the operators will be beneficial, since it is likely that some modifications will be required during production. Certainly, the operators of the FMS, who are likely to be quite few in number, will have to take on far broader responsibilities than would be expected of them in a more traditional manufacturing environment. The training must adequately prepare the operators for these tasks.

As far as the equipment is concerned, all the operators will have to be capable of operating all the equipment. It is likely that each operator will develop preferences but, if the circumstances arise, each must be capable of operating all of the equipment within the FMS. This is quite a demanding task since an FMS is likely to comprise a number of machine-tools with different controllers and tooling requirements, together with robots and material handling systems. However, this does not reflect the true enormity of the task, since not only will the operators have to be able to operate the equipment in the hopefully relatively straightforward, normal circumstances, but more importantly, they must be able to return the FMS to normal operation when an element of it has failed. This is probably where most of the training will be required.

Much of the knowledge necessary for the operators to recover the FMS can be imparted quite effectively during the installation phase of the system. This could, for example, include the ability to exchange workstation controller interface devices. It is operations such as these which emphasise the importance of, wherever possible, using high-quality plugs and sockets to help make the process of installing and testing communications and processing equipment as easy and efficient as possible.

The duration of this entire process will be significantly reduced if an appropriate, well-thought-out plan is established in advance, and supported by well-designed, comprehensive diagnostics. However, it should be borne in mind that there are three phases to problem solving:

(1) Identify the problem
(2) Fix the problem
(3) Minimise the probability of a reoccurrence

The computer control system, if designed correctly, should have the facility to provide assistance for the operators to help diagnose faults. Possibly some specialised equipment will be needed, such as test units to help check correct operation of the communications lines. These test units would be connected into a line in place of a controller interface device. After a diagnostics program had been run, such a device would be able to show precisely what has been received and sent, and therefore assist in the diagnostics process. Of course, this capability could be built into the interface device itself, but if this is suspect, one would hardly want to rely on the results from such a test. It should not be forgotten that, what appear to be relatively insignificant design features can have a considerable impact on the operation of the system. For example, if a group of equipment develops a fault the first action will be to have the attention light switched on. This is fine so long as the bulb itself has not failed. One solution to this would be to fit two bulbs, either both red so at least one is on, or perhaps red and green so that either one or the other should always be on.

In the early stages of the project, well before the recruitment process is started, consideration should be given to the management hierarchy which will eventually be used to operate the plant. One of the key differences between these 'factories of the future' when compared with more traditional plants is that of the multi-disciplinary approach to jobs which must be adopted by operators and management alike, if the system is to be truly successful. Clear consideration should be given to the opportunity to remove layers of management from the facility control hierarchy. Operators and supervisors should be trained to be able to cope with a wide variety of tasks, ideally any that could result (though there are obviously practical limitations to this). There should be no possibility of disputes arising because an operator with the incorrect skill classification carries out a particular task. As mentioned previously, if a job needs to be done, it should be possible for any of the available and close operators to carry out the task.

Inevitably, there will be a period of time after the FMS has been commissioned before full production can reasonably be expected from the system. The duration of this period will be dictated by a number of factors, for example, the quality of the operators, the amount and quality of training they have received, the enthusiasm and commitment displayed by all those responsible for making the plant work, and possibly the number of members of the Project Team who have transferred from design responsibility to operational responsibility. There will be a number of other relevant factors, such as the availability of both the data and material necessary to manufacture the relevant components, not to mention the inherent complexity of the system. Obviously, the duration of this period will vary substantially from system to system, but a reasonable guideline would be from between six to fifteen months.

When trying to estimate this figure, the ambient level of experience of this type of technology should be taken into account. Regardless of the quality of the training that the operators have received, if the operators are very familiar with, say, CNC machines-tools and robots, it would only be the computer aspects of FMS which would need to be learnt. If all aspects of the FMS are new to the factory and/or the operators (a not particularly desirable situation), then the learning process will take considerably longer. However, another issue, the importance of which should not be underestimated, is that of the difficulty which is likely to be experienced in finding people potentially capable of carrying out the required task of operating the FMS.

Obviously the suppliers of the equipment, especially of the computer control systems, will be required to play a significant role in the task of training and starting up the system. This will especially be the case if much of the maintenance of the system is to be carried out using in-house expertise. Certainly, all the purchase contracts must take into account the inevitably lengthy commissioning and learning period, at least by providing an extension to the more typical warranty terms, which are unlikely to cater for the requirements of a complex integrated system where equipment from a number of different manufacturers must interact closely and successfully.

Ultimately it is the FMS Project Team's responsibility to ensure that the operators are adequately trained to operate the system. It might well be one of its last responsibilities with the system (unless they in total or in part are going to become responsible for running the system), but it is certainly one of the most important. Conceivably, this is one of the more difficult tasks if, for example, the technology within the system is new to the company, and indeed that plant itself might be in a new location. There is no doubt that the quality of the operators and their training will exert a significant influence on the eventual success or failure of the system. There is much to be said for a company which takes the trouble to have open days for both customers, families of operators and those who are connected with the system. Hopefully the customers who visit the system will be suitably impressed and give more work. All the others who come will no doubt feel far more comfortable about the technology after they have seen and appreciated the much-talked-about system.

12 Factors Likely to Influence the Development of FMS

12.1 Introduction

As time passes the prospects for FMS, and its counterpart CIM, continue to improve. One cannot argue that FMS has not been through a slow and painful gestation period (not unlike that experienced during the development of numerical controls). However, these times are now all but over. The market pressures for product variety are continuing to force manufacturers in all industries to consider means of producing flexibly. While these pressures have emphasised the need for FMS, manufacturing technology in its turn has evolved to the point where it is feasible to design and implement sophisticated, reliable and efficient flexible manufacturing systems. Probably the only significant issues remaining are:

(1) How can the severe shortage of expertise be overcome?
(2) How can a set of industry-acceptable terms and definitions be established? (FMS, FMC etc. continue to mean significantly different things to different people.)
(3) How can the costs and risks associated with the development of once-off control software be reduced?
(4) How can it be ensured that systems are being designed and implemented well?

However, despite the fact that technology is continuing to advance even more rapidly than before, this does not mean that flexible manufacturing systems will become proportionately easier to implement. Unfortunately, as technology advances, so the requirements become equally more demanding. But the fact remains that FMS is going to become an increasingly more familiar sight within all types of manufacturing environment. It might be given a more grandiose title, perhaps as part of Computer Integrated Manufacturing, but that will not obscure the origin of the technology.

The remainder of this chapter is devoted to a description of how many of the above issues are being addressed, in particular the problems of people, generic control software, communications and some of the more important technologies which currently are being refined.

12.2 The people

Quite rightly, people are expresing a desire not to devote as large a proportion of their day to work as previously. This is particularly the case within a factory environment, regardless of how pleasant that environment has been made (see figures 12.1 and 12.2). Flexible automation now provides an opportunity to satisfy this demand economically.

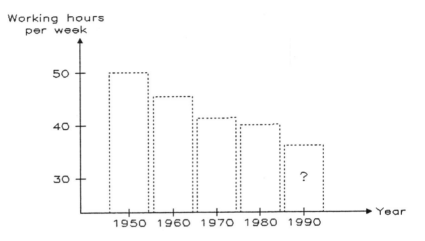

Figure 12.1   Average European weekly working hours in industry

Despite many widely held beliefs, all flexible manufacturing systems of today are operated in 'demanned' rather than an 'unmanned' manner. This is likely to apply even more to the systems of tomorrow. Ideally, these systems operate for twenty-four hours a day, seven days a week, perhaps only stopping briefly for periods of preventative maintenance. Without this level of intense operation it would be difficult to justify the expenditure, especially that which is additional over, say, stand-alone CNC. But the important point is that operators, in fact very highly skilled multi-disciplined operators and maintenance engineers, are required in substantial numbers to run these systems.

At long last, production engineering is being recognised as the highly skilled profession that it is. A profession which is requiring more and more skill and experience, if one wishes to excel, and a profession which can add considerably to the wealth and integrity of a nation.

Universities and Technical Colleges throughout the world have realised that there is a desperate need for a new breed of production engineer, capable of implementing the automation systems of the future. Although it remains important to have a firm grasp of traditional manufacturing techniques, it is

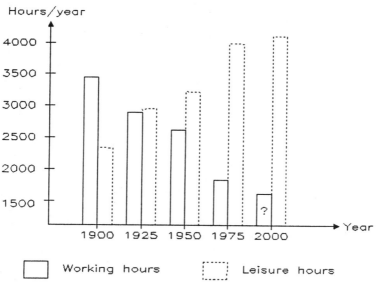

Figure 12.2   As working hours decrease, leisure hours increase

equally important to have a much broader understanding than was previously needed, together with a thorough understanding of the possibilities offered by more relevant automation techniques.

One of the key issues concerning FMS is the multi-disciplinary approach that must be adopted. If someone is a long-time 'expert' in one particular, relatively narrow field, this might actually prove to be a distinct disadvantage. However, having said this, it is important not to forget that working as a member of a team is fundamental to the success of every complex project. Long gone are the days when individual contributions, in total isolation, could turn an idea into a success. Nowadays support is essential, and this means working in teams, optimising the utilisation of complementary skills and experience — an art which, like management, is not so easy as people might like to think.

For example, an 'expert' robot or machine-tool engineer who does not appreciate the problems of interfacing one device to another both mechanically and electrically is going to be of little value in the development of an FMS. Similarly, neither is someone who is unable to act as part of a closely integrated FMS design team.

This problem of how to train people to take maximal advantage of the new manufacturing technologies is now being addressed by many universities and colleges. MSc courses in robotics and even FMS itself are becoming commonplace. However, having the courses is only the first step. Who is going to teach these courses? The learning curve involved in implementing just one

FMS is substantial, and the time needed is anything between three and five years, if not more. FMS is a relatively new technology, and it is debatable whether sufficient time and incentive have accumulated to allow the necessary experience to flow back into academia. All universities and colleges should bear in mind that implementing FMS is very much experience-oriented. The close linking to industry of the teaching of these courses is essential. Certainly, every effort should be made to visit as many systems as is practical, and in particular to meet as many people as possible who have been involved with the implementation of these and other systems.

12.3 Industry-acceptable terms and definitions

One of the most trivial and yet confusing problems associated with the development of any new technology is that of everyone agreeing upon an acceptable set of terms and definitions to describe the technology. This has certainly been a problem which has confronted both FMS and CIM, and it does not look as though the problem is likely to be resolved in the near future.

Even now there is considerable confusion about what is meant by 'flexible manufacturing cell', 'flexible manufacturing system', 'workstation', 'generic controller', and even 'FMS' and 'CIM' themselves. To some degree this is understandable since in different industrial environments these terms, at least initially, are almost bound to mean different things. However, the time during which such a situation might be acceptable is fast ending.

It would certainly be a creditable accomplishment for this book to assist with the generation of an acceptable set of definitions, though not an easy task. However, a Glossary of Terms has been included at the end of the book, which it is hoped will not only aid the understanding of this book, but will also assist generally in the creation of some generally acceptable definitions.

12.4 Generic FMS control systems

One of the most active areas of development concerning FMS and perhaps the one which will have most impact during the mid to late 1980s is the development of 'generic' FMS control systems. These are an attempt to provide an 'off-the-shelf' solution to the control of a wide variety of flexible manufacturing systems. They are essentially an attempt to address two of the problems which continue to confront potential FMS designers:

(1) How to avoid the costs and risks associated with once-off software developments.
(2) How to integrate equipment from a variety of suppliers into one system.

From the suppliers' point of view, generic control systems are an attempt at attracting a very large market to a particular company's products, and in doing so supply the market need for capable and credible single-source co-ordinators for the supply of flexible manufacturing systems. This subject will be discussed in more detail in the following section of this chapter.

However, systems such as these are not the only way of addressing these problems. For example, the advent of communications standards such as MAP and the development of more sophisticated CNC controllers are addressing the issue of equipment integration. But the avoidance of the risks and cost associated with once-off software developments (which typically represent between 15 and 30 per cent of the cost of an FMS) is a more complex and difficult issue, and one which probably is only likely to be resolved by the input of a substantial amount of resources and expertise.

As mentioned above, the main aim of generic FMS control systems developments is to create systems which will be able, with an absolute minimum of customisation, to control many different flexible manufacturing systems. Unfortunately, it is unlikely that a generic controller could be built to meet all possible needs. As is shown in figure 12.3, all that can be hoped is that a generic controller could meet the requirements of a significant number of different systems, to ensure that its development is commercially viable.

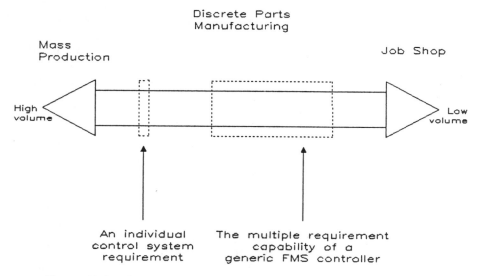

Figure 12.3   Generic controllers within the manufacturing spectrum

There are essentially two approaches to these developments which appear to find general favour. First, to develop a generic software architecture which facilitates configuration to meet the needs of a particular implementation. Second, to provide an integrated total solution which is both generic, within certain bounds, and cost-effective — that is, not needing to penalise the customer for the additional functionality needed to enable the package to control other FMS installations. Both approaches have advantages and disadvantages.

For example, a flexible software architecture approach might well provide a control system for a very wide variety of FMS installations. But a substantial amount of effort is usually needed to actually tailor the software to meet a specific requirement. Then there is the risk that since the software is always being used in slightly different combinations, it is likely to be slightly more unreliable, and require significant debugging.

This approach is likely to be more suited to larger systems, where perhaps cost is not such an issue when compared with the desirability of having the control system tailored almost precisely to the specific installation. To date, however, developments along these lines have not been particularly successful.

Conversely, the 'integrated total solution' is unlikely to be as generic or as expensive as the generic software approach. Also, to be successful, such systems must address the issues of integrating various equipment from different suppliers and providing the appropriate communications support both within the FMS and to other manufacturing or management systems.

Some manufacturers already purport to have such generic control systems on the market — there are examples from West Germany, Japan and the USA. However, careful analysis of these systems shows that they suffer from some fairly significant shortfalls; in particular, they still require a significant amount of application-specific development. However, many companies throughout the world are addressing this issue and no doubt sophisticated, truly generic flexible manufacturing cell controllers will be available soon.

To conclude this discussion of generic controllers, it is worth mentioning generic control at the workstation level. This is essentially the provision of a generic control system for what, in the USA, would be called a relatively small cell, perhaps one machine, one robot and a small stretch of conveyor. Such systems are a useful first step toward both integrated automation and FMS. While they provide some of the advantages of FMS, they are not burdened by the level of complexity typically associated with FMS. If designed correctly, it is quite feasible for these automated workstations to be integrated into an FMS at a later date. At this time, the benefits of work scheduling etc. could be included, without the concerns of taking too large a technical step at one time.

It is likely that these generic workstations will have software which will run on one of three types of computer hardware: on a device controller, perhaps that of a machine-tool or robot, on a dedicated processing device, such as a programmable logic controller, or finally, on the most likely option of a small factory, a ruggedised microcomputer.

Interestingly, systems such as these are already commonplace in process industries, where their main task is one of data collection and analysis. Only recently have the capabilities of such systems been extended to include the requirements of discrete parts manufacturing.

12.5 Single-source co-ordination

Single-source co-ordination is a relatively new term that has found considerable attraction among potential FMS suppliers and purchasers. It is the term being used to supersede that of turnkey supply, a largely outmoded concept in the modern realms of large complex automation systems.

A single-source co-ordinator is still an organisation to whom a potential purchaser approaches for the supply of an FMS. However, the responsibilities of the co-ordinator are essentially to design and implement the FMS, at all times working closely with the purchaser (who might be supplying some of the specialised process technology and probably selecting the process equipment). The main difference from a turnkey supplier is that the co-ordinator is not solely accountable for the design, installation and, in particular, the performance of the total system. While in a turnkey situation, the purchaser may only have notional control over the system being supplied, the single-source co-ordinator shares these responsibilities with the supplier and the customer. This is generally agreed to be a more beneficial relationship to all parties.

This is certainly an important change in the way in which automation systems such as FMS are purchased. During the early 1980s most companies, if they wanted an FMS, went to a contractor who offered a 'turnkey supply' capability. This usually turned out to be a machine-tool company who perhaps saw the selling of 'systems' as the only means of supporting their otherwise declining machine-tool sales.

This was not always particularly satisfactory for the customer. Usually, the systems were, hardly surprisingly, designed with a machine-tool builder's perspective. Frequently it also meant that the customers' choice of equipment was substantially limited to that which the machine-tool builder felt inclined to supply. Most of this, of course, tended to be of their own manufacture. In many respects, this situation led to the implementation of several fairly unsuccessful flexible manufacturing systems which contributed significantly to the generation of the bad reputation of FMS in the past.

Now that the FMS market is more knowledgeable about its own requirements, it has become clear that machine-tool builders are not necessarily the most appropriate organisations for the supply of flexible manufacturing systems. Most customers prefer to select their own process equipment, probably from a variety of manufacturers with whom they are used to working. Ultimately, however, it is neither the material handling equipment nor the machine-tools that in isolation make an FMS successful. It is instead their integration, and most important of all, it is the quality and appropriateness of the FMS control system.

So, how should a good single-source co-ordinator be selected? Unfortunately there is no easy answer to this question. It is relatively straightforward to produce a check-list of the most significant desirable attributes. But it is not so easy to find a company that satisfies them all. The requirements are essentially as follows:

- An experienced core of expertise.
- A sufficiently large and well-qualified support team, for design and implementation.
- Systems-orientated products and/or partnerships.
- Guaranteed market longevity.
- Long-term maintenance capability.
- A reputable company.

An experienced and sufficiently large design and implementation team is essential. Such a team clearly demonstrates an acceptable level of commitment and that a sufficient number of highly qualified people are present, both to handle the quantity of work and to ensure that relevant experience is readily available to guarantee the integrity of the design. Such experience would probably be in mechanical handling, direct or distributed numerical control, local area networks, new production techniques and manufacturing philosophies such as 'Just in Time', not to mention computer control systems etc.

It is ideal if the company concerned, in addition to having a generally good reputation, also has a broad internal manufacturing base. This substantially increases the likelihood of it being able to offer the level of generalised, up-to-date manufacturing expertise which is essential for the design and implementation of successful flexible manufacturing systems.

Although some organisations attempt to develop a substantial business from designing flexible manufacturing and similar systems, it is highly desirable from the customer's point of view that a single-source co-ordinator should have some other interest in the success of the FMS, besides just reputation. Typically, this would tend to be through some form of product involvement. While this is certainly a highly desirable situation, since it does ensure the co-ordinator's continued interest in seeing that the system is successful, care must be taken to ensure that decisions are not unduly biased in any way, perhaps towards the co-ordinator's readily available products.

Certainly, the product involvement could be with the process equipment, but then the situation might be very similar to that of the machine-tool manufacturers, described earlier. Alternatively, and indeed preferably, it should be with the developer of the FMS control system, since it is this which is of most importance to the success of the FMS. This might tend to suggest that software systems houses would be appropriate. This generally is not likely to be the case, largely because it is unlikely that any of these have sufficiently broad manufacturing experience on which to base the equipment design. But perhaps in the not-too-distant future, this also will change.

Obviously, the single-source co-ordinator will be intending to make a profit from the involvement with the project. This could be either directly from the design or through its product involvement, if not a combination of the two. This is certainly something to bear in mind — generally one does get what one pays for, especially if long-term commitment is involved.

There are many consulting organisations which are ready to charge substantial fees for the design of flexible manufacturing and similar systems. Indeed, many have grown significantly as a direct result of government grants primarily aimed at assisting all types of industries to invest in advanced manufacturing technology. Before going to such an organisation, which is unlikely to be able to become directly involved, perhaps via products, in the implementation of the system, there are a number of issues to be considered. For example, as was suggested in chapter 3, one should establish how many of the organisation's previous designs have been implemented. Take the time to visit these sites and see how successful the systems have been. Try to establish how well the implemented design reflects the consultant's proposal — that is, was any additional work required after the design was agreed, and if so, why was it necessary? This, together with contacts with other customers will help establish the long-term credibility and quality of the organisation's work.

While it would be doing a grave disservice to some consultants, it is only fair to point out that on occasion a consulting organisation has designed a system and then left the customer not only with a large bill, but also the task of finding another organisation willing to implement the system. Since there are few, if any, reputable suppliers of FMS who would be willing to take the responsibility for implementing a system which somebody else has designed, the consultant's work could be virtually worthless. At best, the implementer would wish to check a substantial proportion of the analysis, a fairly costly process in itself.

It is also extremely important that the single-source co-ordinator is likely to remain accessible for long-term support, at least for five years after the implementation of the system, and preferably longer. Hopefully this will ensure that a reasonable level of maintenance can be obtained. Currently, a major problem with FMS and similar systems is that, in the long-term, it often proves difficult to obtain support from the supplier, especially for features such as software, where programmers and systems designers are notorious for changing employers every few years. This factor has forced some FMS users to establish their own software resource which is at least capable of maintaining such a system.

If it is possible to identify a single-source co-ordinator who meets the above criteria, then this is the preferable approach for obtaining an FMS. Both customer and co-ordinator should work together to produce an optimal solution to the customer's particular problem. Such single-source co-ordinators are scarce, but they do exist. If one cannot find one, probably the best option is to grow sufficient in-house expertise to at least manage the project.

12.6 Communications

Communications within a factory environment have, to date at least, been a constant source of difficulty. This is not only because the communications systems themselves are not standardised sufficiently to avoid substantial design efforts for each significant application, but also because devices such as machine-tool controllers generally do not possess either appropriate or standardised interfaces. With the likely advent of generic FMS controllers and communications systems such as the Manufacturing Automation Protocol (MAP), these problems are likely to be substantially reduced. Given the increased power and flexibility of the newer device controllers, it might even become possible to eliminate the problem entirely.

Although most readers will probably be familiar with the term 'MAP', perhaps not everyone will be aware of the full implications of this system. Since, in the future, MAP is likely to be an integral part of many manufacturing systems, a brief summary of the history of the development, and the system itself, will be given.

General Motors (GM) has a corporate department called Engineering and Manufacturing Computer Co-ordination which had the responsibility of defining a global strategy for GM for the application of computers to manufacturing environments. It was given formal recognition in November 1980 when it was renamed the Manufacturing Automation Protocol Task Force.

The main purpose of this group was to prepare a specification that would allow common communication among diverse intelligent devices, in a cost-effective and consistent manner. A set of objectives was drawn up:

- To define a MAP message standard which would support application communications.
- To identify applications functions which would be supported by the message format standard.
- Ultimately, to recommend a protocol that would meet the defined functional requirements.

The MAP group published its first major report in late October 1982. It contained some general network and implementational considerations. Since then, additional reports have been published and these have expanded considerably on the comments made originally.

LAYER	TITLE	PURPOSE
7	APPLICATION	Provides all services comprehensible to application programs.
6	PRESENTATION	Transforms data to and from agreed standardised formats.
5	SESSION	Synchronises and manages data.
4	TRANSPORT	Provides reliable, transparent data transfer from end node to end node.
3	NETWORK	Performs message routeing for data transfers between non-adjacent nodes.
2	DATA LINK	Detects errors in messages moved between adjacent nodes.
1	PHYSICAL	Encodes and physically transfers messages between adjacent nodes.

Figure 12.4   The seven layers of the MAP communications system

MAP is based on a number of national and international standards: ISO OSI Model, ISO/NBS protocols and IEEE 802.4 LAN. MAP is defined in seven layers as shown in figure 12.4.

The full MAP implementation with all these layers is a truly comprehensive, if somewhat cumbersome, system. In fact it has turned out to be so complex that, for relatively straightforward applications, it is unnecessarily sophisticated and expensive. To alleviate this drawback, a further MAP system is being defined. It is called 'Mini-MAP' or, more correctly 'MAP EPA' (Enhanced Performance Architecture). Together these systems offer two methods of physically implementing level 1 of the OSI model, namely broadband and carrierband.

The primary difference between these two systems is that in broadband environments, frequency modulation occurs over the many channels; although for the carrierband single-frequency system, signal modulation only occurs over the single channel. EPA offers the opportunity to implement a MAP-compatible communications system without going to the expense of implementing the full MAP system. It includes only a subset of the seven MAP layers, usually layers 1, 2 and 7 (and possibly parts of layers 3 and 6), as defined in figure 12.4, the end result being a somewhat less comprehensive and less expensive system, of higher performance, than the full MAP system.

Although the MAP EPA system is not currently fully defined, and is unlikely to be so before 1988, it is becoming clear which types of environment will eventually use these systems. The full MAP system is likely to appeal to those organisations which require a large number of controlled devices to be communicating with one host computer, perhaps upwards of thirty machine-tools etc. These organisations are probably likely to be the somewhat larger manufacturing organisations and/or those involved in more exotic industries (aerospace, defence etc.). These typically would have a transmission rate requirement of, say, 100–500 megabits per second.

Where the other systems are likely to be used most effectively is within organisations which are keen to adopt a group technology approach to their manufacturing requirements. If manufacturing is carried out within cells or small systems, typically comprising between ten and sixteen workstations, then a MAP EPA network would be ideal for the communications between the devices and, say, the FMS controller. Possibly the cell controller would then communicate with a factory computer system via a full MAP network (see figure 12.5). In these environments, the data transmission rates would be lower, probably between ten and fifty megabits per second.

Certainly, there has been an enthusiastic uptake of MAP in the major industrialised countries. There is already a substantial European following, which has formed a European MAP Users Group. However, there is no denying that the success of MAP may well be due to the lack of viable alternatives, rather than to the excellence of MAP itself. Regardless of this, the mere fact that a substantial equipment-purchasing organisation such as

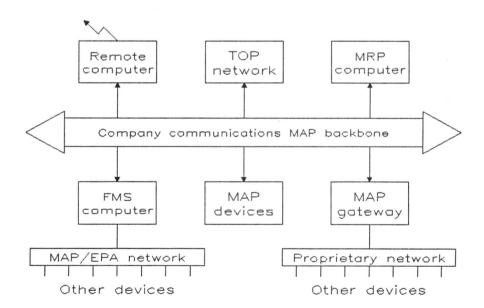

Figure 12.5 A typical MAP communications architecture

General Motors is willing to force its suppliers to comply with the specification is likely to cause it to become a standard one way or another. Most people appear to agree that the long-term effect is likely to be beneficial.

Certainly, companies such as General Electric (USA) have not been slow to show their commitment. The joint venture communications company, Industrial Networks Inc., formed between General Electric (USA) and Ungermann Bass, was the first to announce commercially available MAP hardware. It is now unlikely that any major automation equipment manufacturer will be able to market products incompatible with MAP.

With the involvement of Boeing in the MAP effort, the long-term viability of the effort started by MAP has certainly been enhanced. The Technical and Office Protocol (TOP) is essentially a companion to MAP in as much as it is intended to perform in the office that which MAP is performing in the factory. It assists users who want to exchange information in an office environment comprising computer equipment from many vendors. It provides graphic exchange capabilities, printer and plotter services, electronic mail, file transfer and editable text exchange. While TOP shares some core protocols with MAP, it is likely to be substantially different. This is to be expected since it has different aims and constraints, for example, TOP systems are unlikely to need the same resilience to message corruption as MAP factory floor systems, and hence could possibly use less-expensive, baseband technology as a means of transmission. Nevertheless, while MAP and TOP might need to diverge in the future, every effort is being made to ensure that the two systems remain compatible.

Much has been written about both MAP and TOP, and the interested reader is strongly recommended to consult References and Recommended Reading at the end of the book for some appropriate references.

12.7 Robot technology

Computer simulation of factories has already been mentioned. However, another type of simulation which is beginning to grow in applicability is that of robot simulation.

The aim of robot simulation is to permit a skilled operator to model, usually on a fairly sophisticated CAD system, both a robot and its operating environment. This could be for a proposed application, in which case the system might be helping to establish an optimal selection and layout of equipment. Alternatively, the model might represent an existing application which is to be reprogrammed for a new task.

The benefits of such systems are the possibility of reducing the risk associated with implementing a robotic cell and increasing equipment utilisation by minimising idle time necessitated by reprogramming.

Currently, the main problems with these systems are twofold. First, they are expensive, and second, they are difficult to implement. Although the continual reduction in the cost of computer hardware is having an impact on the first issue (even though software prices are tending to increase), the second issue does not look as though it will be resolved quickly. The result is that, to date and for the foreseeable future, both these systems are likely to be justifiable only within larger organisations where there are a significant number of robots, and where the stopping of one robot results in a substantial loss of production.

The most significant problem within these systems is the accuracy of the model being used to represent the robot's environment. Currently, these models have to be painstakingly created for each specific application, and even then it is debatable how up-to-date these models really are. If the robot has moved slightly, or if some new equipment has been added within the robot's environment, the model would become invalid, and possibly sufficiently inaccurate to result in a catastrophic failure when the program is executed. To ensure such situations do not arise, it is therefore prudent to 'run' the robot program slowly through all the movement combinations, with an operator present to check for any potential problems. These drawbacks can be overcome by, for example, having several fixed datum points strategically placed around the robot's operating area. These can then be used by the robot and the operator to orientate the system. However, currently the need to check programs undermines many of the benefits which might be expected from a robot simulation system. Nevertheless this technology is improving. In years to come, robot simulation is likely to become just as important as automatic part program preparation systems are for machine-tools.

One of the main threats to off-line programming is that it will be superseded by the advances being made in other areas of robot technology, in particular by sensor technology. Many companies already have robots in operation which are purported to be 'intelligent', meaning that they are able to make deductions about their surroundings without the assistance of an operator, or without the necessary logic being specifically programmed into the robot.

Consequently, several robot manufacturers have systems operating which are equipped with tactile sensors. Some are even equipped with force sensors capable of monitoring and adjusting the force being exerted by the gripper. Together these enable the robot to 'feel' and hence grip fragile components with a known force, thus ensuring that the part is not accidently crushed. Such sensors can even enable the orientation of a part to be established. This information could be particularly useful. For example, if a fluorescent light tube was being picked from a conveyor, an appropriate tactile sensor could provide sufficient information to enable the robot to orientate the part reliably and secure it correctly within a light fixture, with the guarantee that the tube would not be broken or incorrectly fitted, regardless of the orientation in which the component arrived. Figure 12.6 is an example of such a system, where a pair of square tactile pads, with their respective ribbon connections, is shown gripping a cylindrical object.

Other 'senses' are also being developed to help produce the intelligent robots of tomorrow. For example, voice input, standardised programming languages with high-level parametric input (that is, referring to points purely by a user-defined name), and, in particular, vision systems. It is likely that during the late 1980s, 3-D vision systems will rapidly supersede the current-technology 2-D systems.

Figure 12.7 shows four pictures of a stack of washers; three of these were obtained with a 3-D vision system. Obviously, considerably more information about the location and orientation of the washers is available using such a system than would be possible with a 2-D system, which might not even be able to separate or identify the three parts.

The incorporation of this technology will produce robots with an awareness of their environment together with the ability to respond automatically to changes in these surroundings. Capabilities such as these will make the robots of tomorrow dramatically different from those of current technology.

12.8 Expert systems

The development of expert systems or artificial intelligence, as it is sometimes called, is a research area of increasing popularity. It is essentially the art of encoding a number of rules which allow a device to ask questions and hence make judgements about the environment in which it is operating. Some of the more publicised systems have been those where a medical doctor's experiences have been encoded, thus enabling a junior doctor, when equipped with an appropriate computer, to diagnose with the skill of a long experienced consultant. Another example is the system developed by General Electric (USA). This facilitates both the diagnosis and repair of faults developed on diesel locomotives.

Factory simulation of FMS has already been covered in depth. However, although the availability and ease of use of computers have been steadily improving, there is no denying that some operator skill is still needed to build and interpret an accurate model. If such a modelling system was to be integrated with an expert system, the result would be a simulation system which, once given the goals of the FMS, could optimise the design almost by itself. Overall requirements would be input, and then numerous iterations would be performed automatically. Interpreted output giving optimal numbers of machines and their layout etc. would be the result. The improvement in the quality of FMS design would probably be substantial.

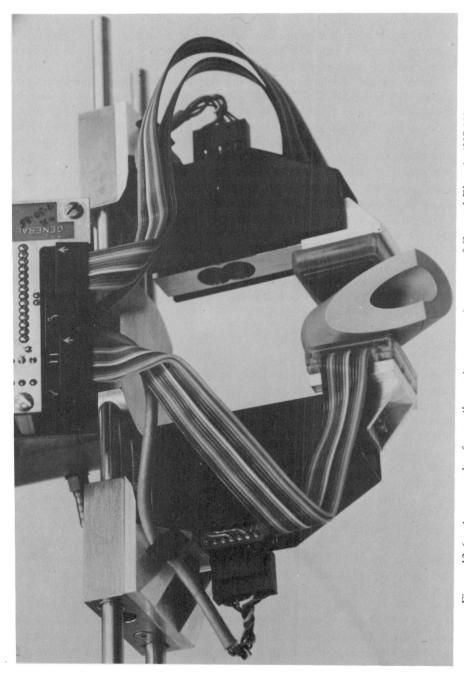

Figure 12.6   An example of a tactile sensing system (courtesy of General Electric (USA))

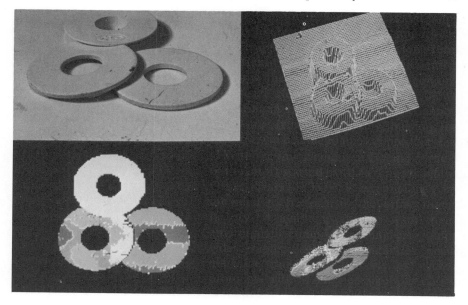

Figure 12.7   An example of a 3-D vision system
(courtesy of General Electric (USA))

The fact that there are so many possible application areas for expert systems, not only within FMS, has promoted the current trend in many research establishments to address the need to develop generic tools to facilitate the 'building' of specific expert systems applications, as opposed to the development of individual applications themselves.

Also, as flexible manufacturing systems continue to increase in their complexity, there is a growing need for assistance in the scheduling of work and the diagnosis of faults. The computer control system within an FMS is, of necessity, monitoring all state changes within the system. If all these state changes are printed out, the ideally paperless environment would soon have more paper than many traditional factories. If a complex error occurs within the system, the only way to diagnose the fault may well be to study all these state changes. One of the problems would be knowing where to start the analysis, since it may not be clear when or why it started. Eventually, if all goes well the origin of the fault would be found; however, this is likely to be both a laborious and skilled task.

If an expert system was built into the control system of the FMS, the origin of the fault is likely to be noticed as soon as it occurs. Such applications would have a substantial impact on the utilisation of an FMS, and hence its economic viability. Certainly, with the advances being made with expert systems, together with the development of generic FMS control systems (thus making the integration of the two technologies more standardised), the possibilities are quite encouraging.

12.9 Manufacturing processes

It is generally true to say that the advances currently being made with the more traditional manufacturing processes themselves have not been occurring either as quickly or in such a far-reaching manner as has been the case with many of the other technologies associated with FMS. However, effort is being expended in order to make, for example, the machining process more effective — better cutting tools are being developed which permit much higher feeds and speeds to be used. Similarly, higher-performance cutting fluids are being developed which not only last longer, but are also more efficient and easier to use than their traditional counterparts. In addition, they can be diluted in different strengths to make them appropriate for more than one manufacturing process. Indeed, some can even be used undiluted as lubricants for the machine itself.

In the past, these traditional process technologies were developed rapidly, largely because it was felt that this was the best, if not the only, way to improve manufacturing efficiency. Now, it is generally appreciated that the proportion of production time that a component actually spends being processed is relatively small, and therefore currently it is the other technologies which are being addressed. So, in searching for more efficient means of manufacturing, higher feeds and speeds are unlikely to be the primary concern. Instead, attention should be focused on other technologies, such as material handling systems and automatic tool change systems, both of which are becoming more popular and sophisticated, as demonstrated by virtually all the major machine-tool manufacturers in all the recent exhibitions. Indeed, not only are machines more frequently being equipped with tool changers but more importantly, many are now being fitted with in-process gauging equipment, tool breakage detection equipment, adaptive machining capabilities, bearing vibration sensors etc.

The emphasis is now being put on more intelligent processing, with equipment which, once started, will be able to monitor and adjust its manufacturing process to maintain continuous and consistent operation within a pre-defined set of process variables.

However, one area which continues to defy progress is that of truly flexible fixturing. Numerous alternatives have been proposed, ranging from a wide variety of traditional modular fixturing systems to the highly innovative phase-change fixturing techniques. Most modular fixturing systems are relatively similar, at least conceptually. They are a clever combination of mechanical parts which provide a means of positively locating a wide variety of differently shaped parts (see figure 12.8). However, no such system has yet been developed which is capable of accommodating all the awkwardly shaped parts which design engineers seem to be able to justify.

Nevertheless, phase-change fixturing does show promise, though further development is still required. The basic principle is to embed a component

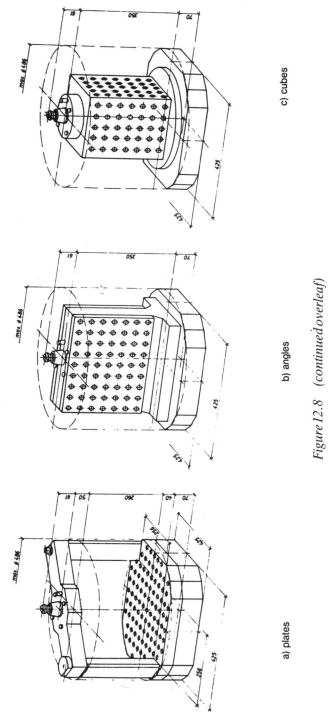

c) cubes

b) angles

a) plates

Figure 12.8 (*continued overleaf*)

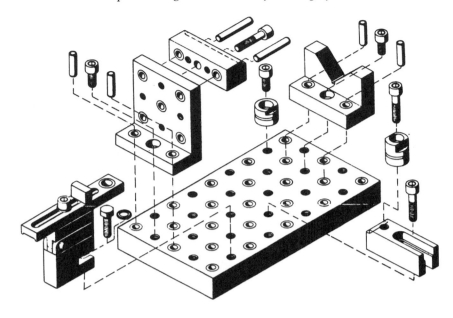

Figure 12.8   Modular fixturing
(courtesy of Blueco-Technik)

within a medium whose nature can be changed from behaving like a fluid, to behaving like a solid. The intention is that the component can be placed both easily and accurately in the fixture medium while it is fluid, and then will be held securely when the phase has been changed. There are a number of systems available. These range from using a container of very small poly-styrene beads, such that when the container is evacuated, from one side, the part becomes securely located, to the use of low melting point alloys which, when allowed to cool and solidify, are able to positively locate a component. In fact, the latter technique has been used for some time to fixture components such as aircraft engine turbine blades (see figure 12.9).

There are some advances which are being made to manufacturing processes which hitherto have not been regarded as being generally applicable to typical traditional production environments. For example, robot-operated systems are now available for both Tungsten Inert Gas (TIG) and Metal Inert Gas (MIG) welding. Many of these are equipped with one of a variety of seam tracking systems, some using voltage detection methods while others rely on laser systems. One of the robotic welding systems, developed by General

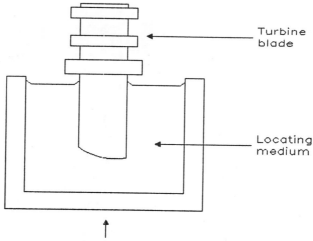

Heat applied to change phase of locating medium

Figure 12.9   Phase-change fixturing

Electric (USA), has a vision system capable of continuously monitoring both the size and geometry of the weld puddle, to guarantee that a near optimum weld is being created. Figure 12.10 shows such a system in operation; the two parallel lines of light guiding the torch can be clearly seen, and also the 'points' of light around the weld puddle which are being used to measure its size.

Another area where substantial advances are being made is that of adaptive control of the spot welding process. Spot welding is a much used process, particularly within the aerospace and automotive industries. Yet until recently, to test whether or not a spot weld was good, the weld had to be destroyed. This led to the situation where typically 20 per cent more welds than actually necessary were made to ensure the integrity of the joint. A system has now been developed, again by General Electric (USA), which is capable of monitoring and adjusting the weld process as it occurs, thus guaranteeing a good joint. A picture of this system is shown in figure 12.11.

Finally, no such discussion would be complete without a mention of the progress which continues to be made with laser and associated technologies

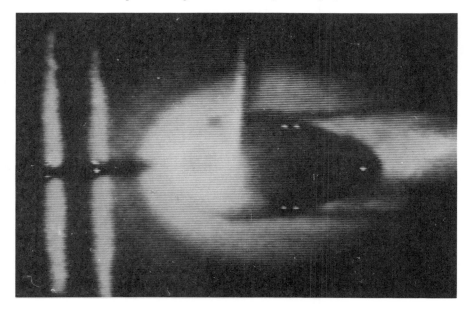

Figure 12.10   The General Electric Weldvision system
(courtesy of General Electric (USA))

within manufacturing. Fibre optic transmission of 400 watt powered lenses is already a relatively well-proven technology and soon 1000 watt laser power will also be transmitted by fibre. Many laser cutting machines are already in service; some 3 kilowatt machines are capable of cutting sheet metal half an inch thick. Indeed, plans are being considered for the development of flexible manufacturing systems based on a central laser power source transmitted, as required, to a number of workstations via a fibre optic link. This emphasises how FMS continues to be a combination of rapidly evolving technologies.

12.10   Computer Integrated Manufacturing

Unfortunately, as mentioned previously, FMS is not the only branch of manufacturing where confusion exists over precise definitions of frequently used terms. This distinction is also shared by Computer Integrated Manufacturing (CIM). A situation which is itself surpassed by the confusion as to where FMS finishes and CIM starts (or is it the other way round?). It will probably be many years before all these technologies have become sufficiently well defined and accepted for the confusion to be ended. However, it does seem to be generally accepted that the basic difference between FMS and CIM is that FMS represents a 'bottom-up' approach to automation, while

*Figure 12.11   An example of a spot-welding system with adaptive control
(courtesy of General Electric (USA))*

CIM represents a 'top-down' approach. Figure 12.12 shows such a hierarchy, which was based on the factory control concept developed by the US National Bureau of Standards.

In the long term, these two approaches will at least tend to merge. Eventually, FMS will probably become just another aspect of any typical sophisticated CIM system. But this is unlikely to occur before the early 1990s, and probably nearer the year 2000. In the meantime, FMS implementers will, no doubt, continue to be concerned with developing sophisticated islands of shop-floor automation, perhaps sharing centralised resources such as automated tool and material storage, transport and computer systems etc. Such approaches allow the benefits of automation to be reaped in stages, with the technological steps involved being made palatable to the user organisation. The key to long-term success is in ensuring that the systems' designs are all part of a long-term automation strategy. This means that the appropriate 'hooks and handles' are available to allow these islands of automation to be integrated in the future into one manufacturing system.

To date, most CIM system developments have started out as the result of a need for some form of Computer Aided Design (CAD) and/or Computer Aided Engineering (CAE), or possibly computer aided process planning. The

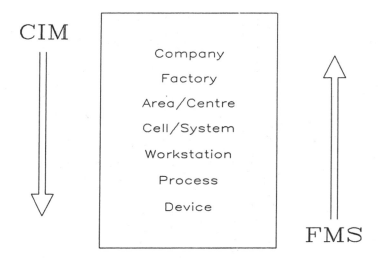

Figure 12.12   The FMS and CIM approaches to automation

many advances that are continually being made with these technologies are assisting with the progress that CIM is making towards more generalised applicability, especially concerning the shop-floor. Currently, however, the cost of entering these computer aided technologies continues to remain high, but this is likely to change quite rapidly as computer hardware becomes less expensive, and as distributed computing becomes more sophisticated.

However, these systems are not generally able to exchange data in a detailed and generic manner. Once truly sophisticated integrated systems that combine many of the capabilities of what would have previously represented individual systems have been developed, then CIM is likely to become a reality. For example, when systems such as computer aided design, computer aided engineering, finite element analysis, mould design and mould flow, solid modelling, simulation, automatic part program generation (for machines and robots), expert systems etc. are all available in a fully integratable form, capable of communicating with, say, FMS controllers, then CIM will have arrived. The eventual users of these systems will be responsible for many aspects of product design and manufacture, from conception of an idea through to control of the manufacture of the part.

Already systems exist which are capable of taking a component design from conception through to production: 'Art' to 'Part' as the technique is sometimes called. General Electric (USA) recently demonstrated its system which allowed a rucksack frame to be designed on a solid modeller (including a finite element analysis to test for strength and lightness). The system was then used to create a die design which was tested for flow and cooling characteristics. Finally, the system was used to generate a machine-tool part program to produce the two die halves. The overall effect was to cut the duration of the design process by several orders of magnitude. Test dies were not required at all, those generated by the system were right first time (see figures 12.13 to 12.15).

It is developments such as these which will form the basis of future CIM systems. FMS is already a good way along its development path. However, over the next few years, the changes which will occur in the general approach to manufacturing will be substantial. Without doubt, FMS as it exists today is the foundation upon which the CIM systems of tomorrow will be built.

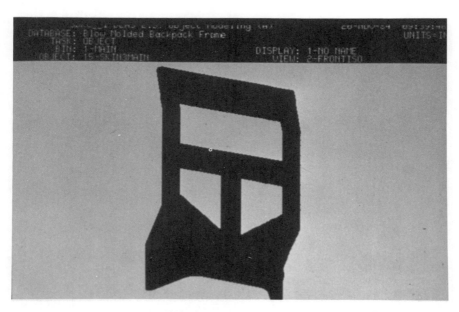

Figure 12.13   The computer model of a rucksack frame
(courtesy of General Electric (USA))

*Figure 12.14 The dies for the rucksack frame generated automatically
(courtesy of General Electric (USA))*

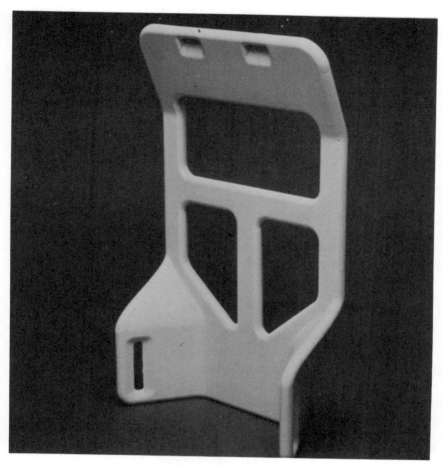

*Figure 12.15   The finish moulded rucksack frame
(courtesy of General Electric (USA))*

References and Recommended Reading

The following references have been used throughout the compilation of this book. In section 1, the papers or articles are listed in sections by chapter purely as an indication of prime relevance. In section 2, a list of books and conference proceedings is included.

(1)   Magazine Articles and Conference papers

Chapter 1

'Automatic manufacture of electronic equipment', *Scientific American*, Vol. 193, No. 2, August 1955.

'Project Tinkertoy', *National Bureau of Standards (USA) Technical Bulletin*, Vol. 37, No. 11, November 1953.

D. T. N. Williamson, 'System 24 — a new concept in manufacture', *8th Int. MTDR Conf., Manchester*, September 1965.

D. T. N. Williamson, 'The anachronistic factory', *Proc. Royal Soc.*, Vol. 331, December 1972.

United States Patent No. 4,369,563, Williamson, 25 January 1983.

J. Browne, D. Dubois, K. Rathmill, S. P. Seithi and K. E. Stecke, 'Classification of flexible manufacturing systems', *FMS Magazine*, April 1984.

K. E. Stecke and J. J. Solberg, 'Loading and control policies for a flexible manufacturing system', *Int. J. Prod. Res.*, Vol. 19, No. 5, 1981, pp. 481–490.

G. T. Farnum, 'FMS: The global perspective', *Manufacturing Engineering*, April 1986.

'Flexible manufacturing, a progression: says Yankee Group', *Automation News*, April 1985.

J. Jablonowski, 'Aiming for flexibility', *American Machinist, Special Report 720*, March 1980.

Modern Machine Shop, 'Not how fast, but how well', February 1985.

K. Kobayashi and H. Inaba, 'The practical approach to unmanned FMS operation', *The Industrial Robot*, March 1982.

Chapter 2

'Big system at John Deere solves new product problems', *FMS Magazine*, Vol. 1, No. 1, October 1982.

I. Nisanci, 'Survey of F.M.S. — applications, problems, & research areas', *CASA–SME FMS 1985, Dallas*, March 1985.

'Study tour of flexible manufacturing systems', *SERC — Napier College, Edinburgh*, October 1982.

S. Birch, 'Rolls Royce takes the lead — just in time', *The Engineer*, February 1986.

H. Challis, 'How does AMT affect the worker?', *Production Engineer*, February 1986.

'Making fridge cabinets in an automated press line', *Production Engineer*, February 1985.

K. Rathmill, 'Japan's FMS complex with laser — a climb down from MUM', *Chartered Mechanical Engineer*, May 1985.

S. Romanini, 'Attaining modularity in systems', *FMS Magazine*, Vol. 2, No. 5, October 1983.

B. Knill, 'Vought Aero Products reveals new lessons in flexible manufacturing', *Material Handling Engineering*, January 1985.

Chapter 3

'Instilling a service mentality: like teaching an elephant to dance', *International Management*, November 1985.

'The factory of the future; science and technology', *The Economist*, April 1986.

R. F. Cota, 'High-Tech products are increasing the engineer's role in marketing', *Machine Design*, March 1985.

S. Seymour, 'Competition analysis', *Chartered Mechanical Engineer*, January 1981.

Chapters 4 and 5

R. Tepsic, 'The parts oriented approach to planning FMS', *FMS Magazine*, Vol. 1, No. 2, January 1983.

S. R. Hayashi, C. E. Thomas, R. K. Davis, W. S. Knight and C. R. Roberts, *Automatic Tool Touch and Breakage Detection in Turning*, General Electric Company, Corporate Research & Development Centre (presented at Hannover Machine Tool Exhibition, September 1985).

J. Jablonowski, 'What's new in machining centres', *American Machinist, Special Report*, February 1984.

J. Hollingum, 'Japan's industry puts its money into FMS', *FMS Magazine*, Vol. 1, No. 2, January 1983.

N. W. Rhea, 'Turbine maker committed to computer integrated manufacturing', *Material Handling Engineering*, January 1985.

A. Dickey, 'Making a tornedo, flexibly', *The Engineer*, August 1985.

'The door has swung open on the age of FMS', *10th Japan International Machine Tool Fair, Metalworking Engineering and Marketing*, January 1981.

R. Baxter, 'Manufacturing system is revolutionary', *Production Engineer*, November 1984.

R. L. Bormann, 'EDM tackles tough jobs', *Manufacturing Engineering*, February 1985.

R. Stokes, 'Current trends in control', *The Engineer*, February 1984.

S. Tulip, 'What's new in metalcutting; Colchester launches high volume turning centre in new strategy', *Production Engineer*, May 1984.

D. Richardson, 'Is metal cutting as efficient as it could be', *Production Engineer*, June 1984.

'The all-in-one flexible turning cell for bar and chucking work', *Production Engineer*, July/August 1982.

A. Postlethwaite, 'Is there any hope left?', *Technology*, October 1983.

K. Stecke and J. Browne, 'Variations in flexible manufacturing systems according to the relevant types of material handling systems', *Material Flow 2*, Elsevier, 1985.

H. Challis, 'Machine tool design "outdated" for unmanned operation', *Production Engineer*, June 1984.

'Focus on materials (ceramics and composite)', *New Technology*, May 1985.

A. L. Wells, 'Computer aided maintenance management', *Computers — Chartered Mechanical Engineer*, November 1984.

M. Orgorek, 'Abrasives technology: Bridging the technology gap', *Manufacturing Engineering*, December 1984.

'Implementing quality control — both internally and externally', *Production Engineer*, February 1985.

'Scanning the horizon (centre drive lathes)', *Modern Machine Shop*, February 1985.

'BP develops new cutting fluids', *FMS Magazine*, Vol. 2, No. 2, April 1984.

M. J. Bragg, 'Lasers make light of industrial jobs', *Chartered Mechanical Engineer*, May 1985.

L. K. Lord, 'Machine tools for use in FMS', *FMS Magazine*, Vol. 2, No. 4, October 1984.

B. Davis, 'Has industry seen the light', *The Engineer*, April 1985.

A. Kochan, 'Advanced cutting fluids for advanced manufacturing technology', *FMS Magazine*, Vol. 2, No. 1, January 1984.

S. Baker, 'Feature — adhesives', *Production Engineer*, March 1985.

Chapter 6

A. Kochan, 'Volvo follows its own transport path', *FMS Magazine*, Vol. 1, No. 2, January 1983.

'Gantry robots bridge the gap in overhead operations', *Robotics World*, September 1985.

G. Hall, 'How can you economically justify a robot purchase?', *Robotics World*, September 1985.

M. Miller, 'Aluminium requires good swarf control', *FMS Magazine*, Vol. 1, No. 1, October 1982.

S. Tulip, 'Three robot users admit their mistakes', *Production Engineer*, July/August 1985.

C. Carberry, 'Robots ease bottleneck in PCB production', *Production Engineer*, March 1986.

R. Macgreggor and T. Janeshutz, 'Playing it safe around robots', *Manufacturing Systems*, September 1985.

A. Cardew, 'Automatic gauging', *Practical Automation*, September 1985.

T. J. Drozda, 'Flexible assembly system features automatic set-up', *Manufacturing Engineering*, December 1984.

A. Kusiak, 'Material handling in flexible manufacturing systems', *Material Flow*, Elsevier Science, 1985.

A. S. Muller and R. G. Hannam, 'Computer aided design using a knowledge base approach and its application to the design of jigs and fixtures', *Proc. Inst. Mech. Eng.,* Vol. 199, No. B4, 1985.

W. F. Nastali, 'Workholding in transition', *Manufacturing Engineering*, February 1986.

'MIG welding. A status report', *Robotics World*, September 1985.

R. T. Wood, L. W. Bauer, J. F. Bedard, B. M. Bernstein, J. Czechowski, M. M. D'Andrea and R. A. Hogle, 'A closed loop system for three-phase resistance spot welding', *Welding Journal*, December 1985.

'When to use lasers, a feature on laser cutting and welding', *Production Engineer*, May 1986.

'Flexible automated assembly recommended', *Robotics World*, September 1985.

'UKL250,000 system increases capacity at Ranco Controls', *Production Engineer*, June 1985.

P. L. Primrose and R. Leonard 'Why robots can be profitable', *Production Engineer*, March 1985.

R. Baxter, 'How 10 robots improved productivity at Llanelli', *Production Engineer*, April 1985.

B. Kraushoff, 'Robot assisted fabrication', *Manufacturing Engineering*, September 1984.

A. Powell, 'Perkins learns a few lessons', *Production Engineer*, May 1985.

A. Baker, 'Automation at Perkins engines', *Chartered Mechanical Engineer*, February 1985.

'The ABC's of X–Y positioning', *Robotics World*, September 1985.

A. Powell, 'Automated assembly boosts output at Antiference', *Production Engineer*, May 1985.

'Assembly: Total systems approach', *Machinery and Production Engineering*, May 1984.

N. W. Rhea, 'MRP II integrated with ASRS', *Material Handling*, February 1985.

L. K. Leng, 'MRP II is the way to do JIT', *Production Engineer*, October 1985.

B. Maskell, 'Just-in-time manufacturing', *Management Accounting*, July 1986.

'Unit load conveyor under divisional control', *Practical Automation*, September 1984.

'Conveyors can cure your handling headaches', *Production Engineer*, November 1985.

P. P. Bose, 'Basics of AGV systems', *American Machinist Special Report 784*, March 1986.

N. W. Rhea, 'Unit load ASRS meets just-in-time manufacturing', *Material Handling Engineering*, November 1984.

Chapter 7

A. Kochan, 'HOCUS takes the risk out of planning', *FMS Magazine*, Vol. 2, No. 2, April 1984.

T. J. Schriber, 'A GPSS/H model for a hypothetical flexible manufacturing system', *Annals of Operations Research*, Vol. 3, 1985, No. 171–188.

K. Rathmill and W. Chan, 'What simulation can do for FMS design and planning', *FMS Magazine*, Vol. 1, No. 3, April 1983.

N. R. Greenwood, R. G. Hannam and B. Valentine, 'Multi machining studies by computer simulation for production planning', *MTDR Conference, Birmingham*, 1980.

J. W. Grant and S. A. Weiner, 'Factors to consider in choosing a graphically animated simulation system', *I.E.*, August 1986.

G. A. Dawson, 'Simulating manufacturing', *Tooling and Production*, December 1984.

'Optimising a manufacturing plant by computer simulation', *CAE, Cleveland, Ohio*, September 1984.

R. R. Duersch and M. A. Laymon, 'Programming-free graphic factory simulation', *Proc. 1985 Winter Simulation Conf. San Francisco*, IEEE 85 CH2214-15.

R. Suri and C. K. Whitney, 'Decision support requirements in flexible manufacturing', *SME Journal of Manufacturing Systems*, Vol. 3, No. 1, 1984.

R. Suri and R. R. Hildebrant, 'Modelling flexible manufacturing systems using mean value analysis', *SME Journal of Manufacturing Systems*, Vol. 3, No. 1, 1984.

K. J. Musselman, 'Computer simulation; a design tool for FMS', *Manufacturing Engineering*, September 1984.

N. R. Greenwood, P. Rao and M. Wisnom, 'Computer simulation for FMS', *IFS SIM — 1 Conference, Stratford on Avon*, March 1985.

Chapter 8

R. B. Bergstrom, 'FMS: The drive toward cells', *Manufacturing Engineering*, August 1985.

P. Wylie, 'L-driver's guide to the shop-floor runabout', *The Engineer*, September 1985.

'Systems and software', *Modern Machine Shop*, February 1985.

D. Willer, 'Designing the computer system for an FMS', *FMS Magazine*, Vol. 2, No. 1, January 1984.

D. Willer, 'Implementing the computer system for an FMS', *FMS Magazine*, Vol. 2, No. 2, April 1984.

R. L. Eshleman, 'A modular approach to DNC/factory management software', *CIM Technology*, Winter 1985.

R. E. Fox, 'Main bottleneck on the factory floor?', *Management Review*, November 1984.

'Let's discuss CAD/CAM integration', *Modern Machine Shop*, February 1985.

J. S. Rhodes, 'The introduction of fully integrated control systems for FMS', *SME FMS Conference*, March 1985.

M. A. Dato, M. Mollo and P. Cigna, 'Modular software for FMS applications', *FMS Magazine*, July 1983.

'What is CAPM and why do we need it?', *Production Engineer*, June 1985.

R. Bonetto, 'Control software at Citroen', *FMS Magazine*, Vol. 1, No. 5, October 1983.

'How to design a system', *Computer Systems*, March 1984.

J. Hollingum, 'Facing up to the programming of machining systems', *FMS Magazine*, Vol. 1, No. 2, January 1983.

A. Wilkins, 'SCAMP comes in modules', *FMS Magazine*, Vol. 1, No. 1, October 1982.

L. Gould, 'Computers run the factory', *Electronics Week*, March 1985.

N. Raman, 'A survey of the literature on production scheduling as it pertains to flexible manufacturing systems', *National Bureau of Standards–GCR–85 499*.

'Machines with minds of their own', *The Economist*, September 1985.

F. E. Harkrider, 'Information management for an FMS', *SME FMS 1985 Conference*, March 1985.

N. G. Odrey and R. Nagel, 'Critical issues in integrating factory automation systems', *CIM Review*, Vol. 2, No. 2, 1986.

R. N. Stouffer, 'General Electric's CIM system automates entire business cycle', *CIM Technology*, Winter 1984.

O. B. Yeoh, F. A. Wilcock and I. T. Franks, 'Monitoring of automated manufacturing systems', *Chartered Mechanical Engineer*, May 1983.

J. Malcolm, *The Design and Implementation of an Integrated Computer Production Control System in a Medium Sized Engineering Company*, UMIST, 1979.

R. Rees, 'Organising for AMT', *Advanced Manufacturing Technology*, July 1985.

R. Haylett, 'OPT — Production control with a difference', *Production Engineer*, May 1986.

J. Jurgenson and L. Altring, 'Information handling, decision making support/generative planning and CIM system integration', *FMS 1984 Conference, Copenhagen*, 1984.

P. Burgam, 'Will CAPP technology replace the planner?', *CIM Technology*, Winter 1984.

J. Hollingum, 'Working towards a common language', *FMS Magazine*, Vol. 1, No. 3, April 1983.

Chapter 9

J. A. Henderson, 'Westinghouse technology modernisation for electronic assembly', *SME FMS 1985 Conference*, March 1985.

'Key to successful D.N.C.', *Modern Machine Shop*, February 1985.

'U.S. MAP users run short of money and patience', *Financial Times Business Information, FinTech 4, 40/1* and *45/7* and *50/7*, 1986.

'SAAB accelerates chassis production with D.N.C.', *Production Engineer*, January 1980.

D. G. Watt, 'Networking software support for F.M.S.', *Manufacturing Engineering*, September 1985.

'Shop-floor controls — can they be standardised', *Production Engineer*, January 1980.

R. Allan, 'Factory communication: MAP promises to pull the pieces together', *Electronic Design*, May 1986.

A. J. Laduzinsky, 'As serial communications buses proliferate, will standards develop? *Control Engineering*, October 1985.

'Traub's with D.N.C. put flexibility into production', *Production Engineer*, July/August 1984.

'One step nearer bolt on D.N.C.', *Production Engineer*, October 1981.

R. G. Hannam, 'Alternatives in the design of flexible manufacturing systems for prismatic parts', *Proc. Inst. Mech. Eng.*, Vol. 199, No. B2, 1985.

Chapters 10 and 11

L. Weaver, 'Investing in advanced manufacturing technology', *Production Engineer*, October 1985.

P. L. Primrose, J. Hoey and R. Leonard, 'A methodology for incorporating the company wide benefits of material requirements planning within a discounted cash flow analysis', *Proc. Inst. Mech. Eng.*, Vol. 199, No. B4, 1985.

V. A. Tipnis and A. C. Misal, 'Economics of flexible manufacturing systems', *SME FMS 1985 Conference* (MS85-154), March 1985.

J. E. Sloggy, 'How to justify the cost of an FMS', *Tooling and Production*, December 1984.

J. Shewchuk, 'Justifying flexible automation', *American Machinist*, October 1984.

P. L. Primrose and R. Leonard, 'The use of a conceptual model to evaluate financially flexible manufacturing system projects', *Proc. Inst. Mech. Eng.*, Vol. 199, No. B1, 1985.

'Adopt automation — but with some consideration', *Production Engineer* (NP 40), May 1984.

'How to justify installing FMS', *Production Engineer*, April 1982.

'Adding up the real costs of factory automation', *Engineer*, April 1986.

H. T. Klahorst, 'How to justify multi-machine systems', *American Machinist*, September 1983.

R. Mills, 'More effective use of NPV in investment appraisal', *Management Accounting*, February 1983.

M. L. Inman, 'Overhead absorption variance analysis — compounding the problem', *Management Accounting*, September 1985.

C. Elphick, 'Cost allocation: a new approach', *Management Accounting*, December 1985.

P. L. Primrose and R. Leonard, 'Conditions under which flexible manufacturing is financially viable', *3rd Int. FMS Conf.*, September 1984.

P. Williams, 'Understanding "just in time" manufacturing', *Production Engineer*, July/August 1985.

M. Jelink and J. D. Goldhar, 'Economics in the factory of the future', *CIM Review*, Vol. 2, No. 2, 1986.

M. Browne, 'Understanding CIM and making it work', *Production Engineer*, December 1985.

A. Palframan, 'A glimpse of the CIM future', *New Technology*, May 1985.

K. W. Nicols, 'Understanding the impact of CIM', *Chartered Mechanical Engineer*, May 1985.

O. B. Davis, 'Renaissance on the factory floor', *High Technology*, May 1985.

(2) Books and Conference Proceedings

The Flexible Manufacturing Technology Market in Europe, Frost and Sullivan, September 1984.

The Flexible Manufacturing Technology Market in the US, Frost and Sullivan, September 1984.

"F.M.S. 1985" Conference Proceedings, March 1985, Dallas, Texas, CASA/SME, March 1985.

"F.M.S. 1986" Conference Proceedings, March 1986, Chicago, Illinois, CASA/SME, March 1986.

MAP Specification, Manufacturing and Engineering Development, General Motors Technical Centre, Warren, Michigan, USA.

Manufacturing Automation Protocol — User's Group Summary, 10–11 September 1985, Annaheim, CA, SME Publications.

MAP Reference Specification, MAP/TOP Users Group. SME, February 1986.

M. C. Bonney and Y. F. Yong, *Robot Safety*, IFS Publications, 1985.

J. R. Holland (Ed.), *Flexible Manufacturing Systems*, SME Publications, 1984.

M. J. Herald and S. Y. Nof, *The Optimal Planning of Computerised Manufacturing Systems*, Purdue University, January 1978.

T. R. Pryor and W. North, *Applying Automated Inspection*, SME Publications, 1985.

Simulation in Manufacturing, Proc. 1st International Conference, IFS Publications, March 1985.

Flexible Manufacturing Systems, Operations Research Models and Applications, Proc. 1st ORSA/TIMS Conference, Ann Arbor, 15–17 August, 1984.

K. Stecke and R. Suri (Eds), *Flexible Manufacturing Systems, Operations Research Models and Applications, Proc. 2nd ORSA/TIMS Conference*, August 1986.

Flexible Manufacturing Systems, Proc. 3rd International Conference, IFS Publications, September 1984.

Flexible Manufacturing Systems, Proc. 4th International Conference, IFS Publications, October 1985.

P. G. Ranky, *The Design and Operation of F.M.S.*, IFS Publications, 1983.

A Competitive Assessment of the US Flexible Manufacturing Systems Industry, Office of Capital Goods, US Department of Commerce, July 1985.

Proceedings — ISATA 1984, Milan, September 1984, Vol. 2.

T. Muller, *Automated Guided Vehicles*, IFS Publications, 1983.

Automated Guided Vehicle Systems, Proc. 16th IPA Conference, Stuttgart, June 1983.

A. Raouf and S. I. Ahmad (Eds), *Recent Developments in FMS, Robotics, CAD/CAM, CIM*, Elsevier, 1985.

M. G. Wright, *Discounted Cash Flow*, McGraw-Hill, 1967.

M. H. Abdelsamad, *A Guide to Capital Expenditure Analysis*, American Management Association, 1973.

R. J. Schonberger, *Japanese Manufacturing Techniques*, Free Press/Collier Macmillan, 1982.

N. L. Hyer (Ed.), *Group Technology at Work*, SME Publications, 1984.

R. W. Yeomans, A. Choudry and P. J. W. Hagen (Eds), *Design Rules for a CIM System*, Elsevier & North-Holland, 1985.

D. Ross (Ed.), *Proceedings of the First International Conference on Machine Control Systems, October, 1985*, IFS Publications.

D. C. Montgomery W. L. Berry (Eds), *Production Planning, Scheduling, and Inventory Control: Concepts, Techniques, and Systems*, American Institute of Industrial Engineers, AIIE-PP&C-74-1.

L. Hammer, K. E. Stecke, R. Suri and J. C. Baltzer (Eds), *Annals of Operations Research — FMS Models and Applications*, 1985.

Glossary of Terms

AGV   Automated Guided Vehicle.

AGVS   Automated Guided Vehicle System.

Area   A term frequently used in the USA to denote a major subdivision of a single factory unit.

ASRS   Automated Storage and Retrieval System, usually a computer-controlled warehouse.

Batch   The number of parts originally associated with an order put on to the shop-floor.

Bay   A logical subsystem of an FMC or FMS. Typically, an area containing several items of process equipment which carry out a particular operation or set of operations on various products. Sometimes called a *cell* or *module* in the USA.

Buffer Store
Temporary storage location (*see* **Work in progress**).

Building
A term frequently used in the USA to denote a single factory unit.

CAD   Computer Aided Design.

CAE   Computer Aided Engineering.

Cell   A small integrated unit of manufacturing equipment. Considerable disagreements exist as to how much equipment (typically 4–8 workstations) is acceptable.

Cell Controller
Small computer device (PC or PLC) used to co-ordinate the activities of a small number of devices.

Centre   A term sometimes used in the USA to denote a major subdivision of an area.

Chips   American equivalent of swarf.

CIM   Computer Integrated Manufacturing.

Cold Standby
A computer back-up capability which requires substantial operator intervention to restart after a computer failure has occurred.

Contingency Management
Non-steady-state occurrences within an FMS, such as start-up, shut-down, equipment failure etc.

CNC   Computer Numerical Control.

Cubic Components
Parts which do not display rotational symmetry. Typically manufactured on a machining centre.

DCF Discounted Cash Flow; a financial analysis technique for appraising capital projects.

Debug Removal of final problems (bugs) from a system.

Device An item of manufacturing equipment (such as a machine tool, a robot etc.), possibly equipped with a device controller.

Device Controller
Typically, a type of computer used to control an item of manufacturing equipment (such as a machine tool controller, robot controller etc.).

Discrete Parts
Individual parts; these could be manufactured in either job-shop or mass-production environments, depending on the volume required (*see* **Process Industry**).

DNC Direct Numerical Control; a communictions system between CNCs and a host computer allowing upload and download of part programs, and sometimes machine status.

EPA Enhanced Performance Architecture; a subset of the MAP communications system designed to be less functional, of higher performance and less expensive to implement.

FMC Flexible Manufacturing Cell (usually a small FMS, comprising only up to 16 workstations); dividing lines between FMC and FMS are ill-defined.

FMS Flexible Manufacturing System — usually a large FMC, comprising more than 16 workstations (*see* Chapter 1).

Host Computer
Computer responsible for controlling the FMS.

Hot Standby
A computer control system typically comprising two computer units, one of which automatically continues to operate the FMS in the event of the other failing.

Icon A symbol used on a computer screen to represent an item of equipment. Typically as part of a representation of the entire FMS on a mimic screen.

IRR Internal Rate of Return; financial measure for the viability of an investment.

JIT Just In Time (the philosophy of producing parts only when they are actually needed).

Job Shop
A small manufacturing facility which generally makes small numbers of a wide variety of different parts for various, often large, customers. The archetypal small batch manufacturer.

Live Computer

> Computer currently responsible for controlling the FMS, as opposed to the standby.

Lot The number of parts typically transferred on a pallet (that is, a subdivision of a batch).

MAP Manufacturing Automation Protocol; protocol being developed to facilitate multi-vendor equipment communications on the shopfloor (*refer to* **TOP**).

Mass Production

> The production of discrete parts in large volumes, usually with little or no variety.

MIG Metal Inert Gas welding.

MIMIC Computer display representing the FMS, usually showing status real-time information.

Module Term denoting an integral part of the whole; could be applied to an FMS which might comprise several modules (or subsystems), or to the FMS control software.

NPV Net Present Value.

Off-line Programming

> The programming of a device on a remote computer, or at least in such a way as not to prevent the device from being productive.

Operation

> An individual step in the manufacturing process for a particular product, typically carried out on one item of production equipment.

Order

> An order represents a request for a manufacturing system to produce a batch of a particular product, usually including delivery data and other related information.

PC Personal Computer.

PCB Printed Circuit Board.

PLC Programmable Logic Controller (sometimes called a *programmable controller* — not to be confused with a PC).

Process

> A process is the sum of the operations performed within a particular bay.

Process Industry

> The production of continuous products, such as chemicals etc., essentially the opposite of discrete parts manufacture.

Recovery

> The return of an element of an FMS to normal operation after a failure (*see* **Contingency Management**).

ROI Return On Investment.

Rotational Components

> Components which display a rotational axis of symmetry (*see*

Turned Components).

Set-up   The work needed to change equipment from having been able to produce one type of product to being able to produce another.

Single Source Co-ordinator
Supplier who, while sharing overall risks and responsibilities with the user, co-ordinates the implementation of the entire FMS.

Solid Modelling
The use of a CAD system to generate models which contain details relating to all points on the surface of and within an object.

Storage Area
An area in which parts are stored — could be purely floor space, could be an ASRS.

Subsystem
A logical subdivision of an FMS (*see* **Bay** *and* **Module**).

Swarf   Waste material typically generated during a machining operation.

TIG   Tungsten Inert Gas welding.

TOP   Technical and Office Protocols; protocols being developed to facilitate communication between multi-vendor computer equipment in an office environment (*refer to* **MAP**).

Transportation System
A material handling system usually associated with the task of transporting lots of parts between bays.

Turned Components
Rotational components produced on a lathe.

Turnkey Installation
Facility where one supplier has been solely responsible for the implementation of the entire FMS.

Warm Standby
A computer back-up system which operates after a minimal amount of operator intervention when a failure has occurred.

Work in Progress
Material which has started in production which cannot yet be classified as finished parts.

Workstation
An individual processing unit. Usually a sophisticated machine tool with ancillary equipment. Sometimes used to define a small number of closely integrated production devices, typically a machine tool, a robot and a small length of conveyor.

WSIU   Workstation Interface Unit; a device used to connect device controllers to an FMS communications network and provide an operator terminal close to the process equipment.

Index